Wing-Level Mission Assurance for a Cyber-Contested Environment

DON SNYDER, LAUREN A. MAYER, JONATHAN LEE BROSMER,
ELIZABETH BODINE-BARON, QUENTIN E. HODGSON, MYRON HURA,
JONATHAN FUJIWARA, THOMAS HAMILTON

Prepared for the Department of the Air Force
Approved for public release; distribution unlimited

T0308350

PROJECT AIR FORCE

For more information on this publication, visit **www.rand.org/t/RRA580-1**.

About RAND

The RAND Corporation is a research organization that develops solutions to public policy challenges to help make communities throughout the world safer and more secure, healthier and more prosperous. RAND is nonprofit, nonpartisan, and committed to the public interest. To learn more about RAND, visit www.rand.org.

Research Integrity

Our mission to help improve policy and decisionmaking through research and analysis is enabled through our core values of quality and objectivity and our unwavering commitment to the highest level of integrity and ethical behavior. To help ensure our research and analysis are rigorous, objective, and nonpartisan, we subject our research publications to a robust and exacting quality-assurance process; avoid both the appearance and reality of financial and other conflicts of interest through staff training, project screening, and a policy of mandatory disclosure; and pursue transparency in our research engagements through our commitment to the open publication of our research findings and recommendations, disclosure of the source of funding of published research, and policies to ensure intellectual independence. For more information, visit www.rand.org/about/principles.

RAND's publications do not necessarily reflect the opinions of its research clients and sponsors.

Published by the RAND Corporation, Santa Monica, Calif.
© 2021 RAND Corporation
RAND® is a registered trademark.

Library of Congress Cataloging-in-Publication Data is available for this publication.
ISBN: 978-1-9774-0792-4

Cover: U.S. Air National Guard photo by Senior Master Sgt. John Rohrer.

About This Report

The objective of this project was to identify ways to improve mission assurance at the operational level (i.e., wings, Air Operations Centers, and deltas) in a contested cyber environment by giving recommendations for wing-level commanders and Mission Defense Teams (MDTs) and recommendations for other organizations to better support efforts at the wing level. It builds on previous RAND work.[1] It should be of interest to the operational and cybersecurity communities in the Department of the Air Force.

The research reported here was sponsored by the Deputy Director of Operations, Air Combat Command and conducted within the Force Modernization and Employment Program of RAND Project AIR FORCE as part of a fiscal year 2020 project, "Resiliency to Aircraft Cyber Threats."

RAND Project AIR FORCE

RAND Project AIR FORCE (PAF), a division of the RAND Corporation, is the Department of the Air Force's (DAF's) federally funded research and development center for studies and analyses, supporting both the United States Air Force and the United States Space Force. PAF provides the DAF with independent analyses of policy alternatives affecting the development, employment, combat readiness, and support of current and future air, space, and cyber forces. Research is conducted in four programs: Strategy and Doctrine; Force Modernization and Employment; Resource Management; and Workforce, Development, and Health. The research reported here was prepared under contract FA7014-16-D-1000.

Additional information about PAF is available on our website: www.rand.org/paf/

This report documents work originally shared with the DAF on September 9, 2020. The draft report, issued on September 29, 2020, was reviewed by formal peer reviewers and DAF subject-matter experts.

---

[1] Hagen et al., 2021.

Contents

Figures

Tables

Summary

Issue

Wing-level organizations need to assure their mission(s) despite adversary cyber operations. Current initiatives to empower wings to this end are in their infancy. We address ways in which a wing can advance mission assurance in the face of cyber attacks along four strategic lines of effort:

- to defend its systems
- to respond to and recover from cyber incidents
- to maintain resiliency of its missions when systems fail
- to maintain sufficient situational awareness to make decisions to accomplish defense, response and recovery, and resiliency.

In this report, we step through tasks for each of these strategies and identify current deficiencies in each task group along with potential remedies. Not every task for each of these strategies can be done at the wing level. But the wing will play key roles in each. One of the topics we discuss is which roles should be done at the wing level and which should be done elsewhere, supporting the wing.

Approach

We conducted a range of interviews to discover the current state of affairs, identify the most important issues to resolve, and elicit ideas for improvement. We also examined the literature on cybersecurity and extracted principles from organizational theory to recommend improvements.

Recommendations

Wing-Level Recommendations

Foremost, we recommend that wing commanders take full ownership of cyber mission assurance of the wing. They must see such resources as Mission Defense Teams (MDTs) as tools at their disposal and use them as needed for the wing's cyber mission assurance. They must also understand that MDTs alone will not provide cyber mission assurance. Every member of the wing and every organizational unit within the wing must play a role.

Therefore, we recommend that wing commanders issue a commander's intent regarding cyber mission assurance of the wing. As part of this command direction, we recommend that wing commanders

- create teams for the ability to survive and operate in a cyber-contested environment who

- perform and maintain mission maps for the wing using Functional Mission Analysis-Cyber (FMA-C) or a similar method
- recommend preemptive adjustments to mission architecture for enhanced resiliency
- devise and exercise plans for mission continuity during and after cyber attacks
- advise the commander, during and after cyber attacks, of potential courses of action for mission continuity.

- use MDTs exclusively for cyber defense of systems, so that they
  - defend systems, such that some team defends each system type
  - respond to cyber attacks of systems and recover system capabilities thereafter.

We further recommend that wing commanders

- create an appropriate learning culture for cyber mission assurance to solve the many problems in this area that do not have formulaic answers
- develop squadron and group commanders to have better cyber mission assurance expertise so that the next generation of wing commanders possesses better heuristics.

Above-Wing-Level Recommendations

Major Commands and Field Commands

We recommend that Major Commands (MAJCOMs) and Field Commands establish more-centralized command and control for the response to and recovery from cyber incidents, including the following:

- clear thresholds for what constitutes a reportable incident and processes for reporting
- a command center to prioritize response, assign incidents to organizations for triage and assessment, and disseminate and compile findings.

The command center should be

- able to handle command and control for cyber incidents for information technology systems, industrial control systems, and cyber-physical weapon systems
- operational 24 hours per day, able to surge to wartime needs, and able to handle compartmented and special-access information.

We also recommend that MAJCOMs and Field Commands establish accountability for wings and deltas to evaluate wing and delta commanders on their readiness with respect to the commander's intent issued by MAJCOMs and Field Commanders (see the next section). We recommend that this accountability cover

- administrative compliance: Are the directed actions being taken as envisioned?
- operational effectiveness: Are the actions taken by the wing achieving, or likely to achieve, the goal? This component should be included in operational readiness and other inspections to give it the prominence it needs for wing-level readiness.

To facilitate the quality and use of FMA-C products, we recommend that MAJCOMs and Field Commands

- provide quality assurance checks of finished FMA-C products that wings perform
- create training for wing commanders on how they can best use FMA-C products.

We further recommend that MAJCOM and Field Command commanders issue a commander's intent to wing commanders along the lines of directing them

- to defend the wing's critical systems
- to plan and exercise the ability to respond to and recover from cyber incidents
- to plan and exercise means to maintain resiliency of the wing's missions by adaptation when systems fail
- to maintain sufficient situational awareness to make decisions to accomplish defense, response and recovery, and resiliency.

Program Management Offices

We recommend that Program Management Offices

- develop the appropriate tools for cyber defense of cyber-physical weapon systems
- provide technical guidance to MDTs for how cyber defense of cyber-physical weapon systems ought to be performed, perhaps in the form of technical orders
- develop training for MDTs tailored to their cyber-physical weapon system.

Air University

We recommend that Air University

- enrich FMA-C training to help students move from simple classroom examples to the complicated missions of a wing
- provide exemplars or templates of sound FMA-C analysis that are similar to a wing's mission
- expand, temporarily, to provide either reachback capabilities or field teams to assist wings when doing mission mapping.

Acknowledgments

We thank Ted Uchida for sponsoring this project and for his support throughout. Col Jeffrey Phillips and Lt Col Jarrod Norris were supportive action officers, without whom this work could not have been done. Lt Col Emmanuel (Manny) Matos also helped keep us current on the evolving initiatives, fielded many questions, and provided insights into our recommendations. A key part of this project was dozens of detailed interviews with wing-level personnel, including wing commanders. Those interviews fell during a critical time early in the crisis response to the coronavirus pandemic. Despite the burdens of responding to the pandemic and maintaining readiness, the wings generously gave us their time and insights—a testament to how important mission assurance in a cyber-contested environment is to these units and personnel. We thank them for both their time and their commitment to this aspect of the mission.

We thank Jared Ettinger, Sandra Evans, Gen (Ret.) Donald Hoffman, Drew Lohn, and Laura Werber on the formal review team for their constructive insights and recommendations. Their feedback helped shape the interview protocols and improve the final report. During his fellowship year with us, Lt Col Dennis Borrman helped us considerably with insights, fielded our many questions, and served as a test subject for our interview protocols. John Drew was a source of insight on wing-level operations. We are grateful to Michelle Horner for her able assistance in setting up the interviews.

That we received help and insights from those acknowledged above should not be taken to imply that they concur with the views expressed in this report. We alone are responsible for the content, including any errors or oversights.

Abbreviations

616 OC	616th Operations Center
ACC	Air Combat Command
AFGSC	Air Force Global Strike Command
AFRC	Air Force Reserve Command
AFSC	Air Force Specialty Code
AMC	Air Mobility Command
AO	Authorizing Official
AOC	Air Operations Center
C2	command and control
CC	commander
CCIR	commander's critical information requirement
CDCC	Cyberspace Defense Correlation Cell
CDOC	Cyberspace Defense Operations Center
CHOP	change in operational (control)
CIO	Chief Information Officer
CPT	Cyberspace Protection Team
CROWS	Cyber Resiliency Office for Weapon Systems
CVA/H	Cyberspace Vulnerability Assessment/Hunter
DoD	U.S. Department of Defense
EITaaS	Enterprise Information Technology as a Service
ELICSAR	Enterprise Logging Ingest Cyber Situational Awareness Refinery
EW	electronic warfare
FMA-C	Functional Mission Analysis-Cyber
ICBM	intercontinental ballistic missile
ICS	industrial control system
IS	Intelligence Squadron

ISR	intelligence, surveillance, and reconnaissance
ISRD	Intelligence Surveillance Reconnaissance Division
IT	information technology
MAJCOM	Major Command
MDT	Mission Defense Team
NASIC	National Air and Space Intelligence Center
NIPRNet	Non-Secure Internet Protocol Router Network
OODA	observe, orient, decide, and act
OSS/IN	Operational Support Squadron/Intelligence
PACAF	Pacific Air Forces
PIR	prioritized information requirement
PMO	Program Management Office
RFI	request for information
R&R	roles and responsibilities
SIPRNet	Secure Internet Protocol Router Network
SQS	squadron
USAFE	U.S. Air Forces in Europe
USSF	U.S. Space Force
UTC	unit type code
WCC	wing commander

1. Background

Stating the challenge clearly is half the battle. Scientists and engineers, especially, are prone to get bogged down in explorations of their phenomenologies. They need clear direction to focus on a specific operational problem.

— Glenn A. Kent[2]

Systems and operations have become more and more dependent on interconnected electronics and data. As that dependency grows, so too does the need for mission assurance in the face of adversarial operations through cyberspace.[3] This increasing threat has been well described, and every part of the Department of the Air Force has roles to play in meeting this threat.[4] A key nexus for mission assurance lies at the organizational level where operational and support activities come together to produce Department of the Air Force missions—the wing level.

The Goal

The ultimate goal is mission assurance. A commander, for example, sets a goal of accomplishing some military mission. The commander does not set the goal as suffering no casualties or equipment loss. Minimizing casualties and equipment loss is a constraint. The commander needs to understand how many casualties and how much equipment loss can be absorbed and yet still accomplish the mission. Operations in a cyber-contested environment share this goal of mission assurance with the constraints of minimizing damage.

Therefore, in this context, the goals of wing-level organizations are to limit adversary exfiltration through cyberspace of information in the wing's care below an acceptable threshold and to ensure an acceptable level of operational functionality of the wing's mission(s) even when the wing is attacked offensively through cyberspace. Offensive cyber attacks include degradation of communications and the corruption or destruction of information. We use the term *cyberspace operations* to include both exfiltration of information through cyberspace and cyber attacks.[5]

---

[2] Kent et al., 2008, p. 163.

[3] We use the term *mission assurance* to mean "measures required to accomplish essential objectives of missions in a contested environment" (U.S. Air Force Doctrine, 2020).

[4] Selected key reports and books include Gosler and Von Thaer, 2013; Danzig, 2014; U.S. Government Accountability Office, 2018; Sanger, 2018; King and Gallagher, 2020; and Buchanan, 2020.

[5] See Snyder et al., 2015, p. 2; and Committee on National Security Systems, 2015.

Exfiltration of information can have short- and long-term negative effects on a wing's ability to carry out its missions. In the short term, Red can use exfiltrated tactical information to improve its situational awareness and get inside Blue's observe, orient, decide, and act (OODA) loop.[6] In the long term, Red can use information about weapon systems and their tactical employment to accelerate the fielding of new weapon systems and develop tactics to counter Blue's weapon systems.

Offensive cyber attacks can disrupt mission operations in various ways. Although the vector of attack travels through cyberspace, cyber attacks can directly damage or destroy systems beyond cyberspace, as was done in the Stuxnet operation.[7] They can also disrupt processes by taking control of automation, interrupting communications, or altering or deleting data. And they can confuse decisionmaking and severely interfere with task execution by all or any combinations of these means, as happened in the aftermath of the NotPetya malware infections.[8]

To combat adversarial cyberspace operations, a wing-level organization has three lines of effort to wield:

- It must defend against exfiltration and attacks, which we call *cyber defense.*
- It must be able to restore system functionality after a cyber attack, which we call *cyber response and recovery.*
- It must develop and exercise plans to survive and operate during and after cyber attacks, before recovery has occurred, which we call *cyber resiliency.* Cyber resiliency is shorthand for mission resiliency in the face of cyber attack. Cyber resiliency can be achieved by means outside cyberspace, by using alternative systems and processes to accomplish mission tasks. All members of the wing play a role.

In the cat-and-mouse game of cyber operations, neither cyber defense nor measures for cyber resiliency can be perfect. Hence, there is also a fourth line of effort that is not cleanly separable from the other three:

- The organization must understand what constitutes acceptable levels of cyber exfiltration and cyber attack, how exposed its missions are to reaching these thresholds, and how well it is ensuring its mission against adversary cyber operations, which we collectively call *cyber situational awareness.*

Key questions for cyber situational awareness include the following: In the architecture of the mission, which elements are most critical? Where do individual systems play roles in mission execution, and how critical are those systems, given those roles? How susceptible are the data and systems to cyberspace operations? In what ways are the missions susceptible to cyber operations?

---

[6] The OODA loop concept was developed by John Boyd. The central idea is that the player, Red or Blue, that can observe, orient, decide, and act faster than the other can gain the initiative and keep the adversary confused. Boyd never formally published his work; for a fuller elaboration, see Freedman, 2013, pp. 196–201.

[7] Langner, 2013.

[8] Greenberg, 2018.

Although the term is imperfect, we will call these four lines of effort wing-level *cyber mission assurance*. The phrase is shorthand for *mission assurance in a contested cyber environment at the wing level*. We emphasize that these four lines of effort involve much more than monitoring and defending systems. A theme throughout this report will be that cyber mission assurance is not exclusively a cybersecurity problem. It is a mission problem. Everyone with some role in the mission of the wing-level organization has a role in all four lines of cyber mission assurance.

The challenges for cyber mission assurance have no reprieve in time or space. A wing needs to maintain its cyber mission assurance during peacetime and wartime. But it must recognize that success in peacetime does not necessarily mean that it will have success in wartime. Wartime can confront a wing with adversarial cyber operations of higher sophistication than encountered during peacetime, requiring greater skills to respond, and with a higher volume of attacks against a wider range of targets, requiring greater capacity to respond. Successful strategies must work at home station as well as when deployed, for those units that deploy to a location that is different from home station.

There are, of course, limits to what a wing-level organization can achieve organically. It also needs support from other organizations. That support comes in many forms, including supplying adequately trained personnel, providing tactical cyberspace intelligence assessments, performing cyber forensics analysis, and augmenting its capabilities with other technical assistance.

The purpose of this report is to recommend workable, effective strategies for how the Department of the Air Force can better organize, train, and equip to achieve wing-level cyber mission assurance, with an eye toward the best use and constitution of Mission Defense Teams (MDTs).

Wing-Level Organizations

Wing-level organizations carry out a variety of missions and face diverse challenges for cyber mission assurance. For the purposes of this report, we use the term *wing-level organizations*, or sometimes the abbreviated term *wings*, to mean wings and Air Operations Centers (AOCs) in the U.S. Air Force and deltas in the U.S. Space Force.[9] There are other organizations in the Department of the Air Force at a comparable level, such as Program Management Offices (PMOs) led by a senior materiel leader. They, too, need to continue their assigned missions in the face of adversary cyber operations. But we do not specifically address these units because they have significantly different missions, are organized differently than the wing-level organizations that we treat, and, as of 2020, do not have MDTs.

Whether a wing (or delta) serves as a host unit or is a tenant unit is important for how it ensures cyber mission assurance. Critical systems that support a wing's mission could include

---

[9] Kirby, 2020.

those that provide power, communications links, and other infrastructure under the responsibility of the host unit. Host and tenant units will need to collaborate effectively to see the mission space holistically to ensure the missions of all units on an installation. And because a tenant unit will not have organic expertise in infrastructure support, the U.S. Air Force will need to ensure a deployable capability that can seamlessly integrate with elements of a tenant unit when it needs to deploy.

Wings perform diverse missions with an assortment of weapon systems and other supporting systems. The primary weapon systems for many wings are aircraft. Some aircraft-oriented wings—such as those with combat aircraft—organize, train, and equip to supply forces, when required, to combatant commanders. Day to day at home station, they do not generally perform operations in support of a combatant commander. When combat aircraft support a combatant commander, they do so under the operational control of that commander as an Air Expeditionary Wing.

Other aircraft-oriented wings—such as mobility aircraft and intelligence, surveillance, and reconnaissance (ISR) aircraft—organize, train, and equip and simultaneously support combatant commanders. For these reasons, the needs for cyber mission assurance support during deployment will generally vary across these mission types. As combat aircraft undergo a change in operational control (CHOP) to a geographic combatant commander, they need a deployable unit type code (UTC) capability for cyber mission assurance that does not exhaust the capabilities at home station. Mobility aircraft can both be CHOPed to a theater and fly en-route structures and therefore need to be able to support both. ISR aircraft sometimes deploy, but they do not CHOP to a theater.

A number of wings do not operate aircraft. Some operate information technology (IT)–based networks, such as the Distributed Common Ground System for intelligence. Others operate intercontinental ballistic missiles (ICBMs). Deltas operate space systems, both in the ground segment and the space (orbital) segment. These wing-level organizations deploy in place and do not require UTCs.

The primary mission of an AOC is to produce and execute an air tasking order. To do so, the AOC uses the Falconer weapon system and a variety of other systems. Most of these are networked IT systems using internet protocols. Each AOC operates out of a default location but must be able to operate from alternate locations if under attack. So, it has a need to be able to shift operations geographically, but not in the sense of deployment of other wing-level units.[10]

Some of these wing-level organizations own their missions. A wing that flies fighters prepares to present forces to a combatant commander. The wing commander owns that organize, train, and equip mission.[11] The commander has the authority, for example, to ground the aircraft

---

[10] Air Force Instruction 13-1AOC, November 2, 2011, Incorporating Change 1, May 18, 2012.

[11] Many of the organize, train, and equip authorities lie above the wing level. Relevant to this report, wing-level commanders have the authority to organize the MDT as they see fit, including its placement in the wing. They also

and temporarily suspend wing operations. A wing that operates ISR aircraft or an ISR networked weapon system, however, often supports missions that are owned by some other entity. This distinction is important when we discuss the span of authority and decisions of a wing commander. ISR and similar wing commanders can make certain operational decisions only in collaboration with mission owners. These distinctions in span of control determine key aspects of how a wing should approach cyber mission assurance. We discuss these aspects in Chapter 3.

The systems that support these missions vary across the full spectrum of how cyber systems are generally categorized. Some, such as aircraft and missiles, are what are called *cyber-physical systems*, which are systems in which the processors and software are integrated with the hardware in a way that renders them functionally inseparable.[12] Cyber-physical systems often do not use internet protocols, these systems sometimes run bespoke operating systems and software, and software often runs in real time, synchronized with and controlling hardware states. Failure of properly integrated software and hardware states can lead to system failures and safety hazards.[13]

Some systems that provide power, fuel, traffic, and other utilities that are needed to support missions are most frequently called *industrial control systems* (ICS).[14] These are also specialized systems, but they are often networked using internet protocols for remote control. Because they control power, fuel, traffic, and other utilities, they are foundational to nearly every other capability.

Lastly, some systems, such as the Distributed Common Ground System and the systems used in the AOC weapon system, are internet protocol–based systems that are generally referred to as *IT*. A wing can have any or all three of these categories of systems.

Each wing, therefore, has distinct needs for cyber defense of distinct systems; cyber mission assurance of a variety of missions, some of which the wing might not own; and cyber situational awareness of a distinct terrain.

This report will focus on the wing-level cyber mission assurance of weapon systems. Those weapon systems include cyber-physical systems, ICS, and IT systems. IT systems that are not weapon systems are out of the scope of this report.

---

have some latitude in assigning who has what cyber mission assurance roles and responsibilities within the wing and how the wing trains for cyber mission assurance. They, of course, must conform to any higher-level direction and must pass requisite inspections. As leaders, they set the tone of the culture in the wing with regard to cyber mission assurance.

[12] Weapon systems that are cyber-physical systems are also sometimes called *platform IT* and *embedded software*, although these terms are not always used in ways that are fully interchangeable. See Greer et al., 2019.

[13] Alur, 2015.

[14] What we call *ICS* throughout this document goes by several names, including *operational technology* and *control systems*. See HQ USAF/A4, 2019. Occasionally in the literature, ICS is grouped with cyber-physical systems (Ashibani and Mahmoud, 2017).

Key Supporting Elements

Being embedded in a much larger enterprise, the wing is not alone in its cyber mission assurance responsibilities. In numerous ways, various other organizations help wings. For the Department of the Air Force to fully address the problem of cyber mission assurance, all of these entities need to be working toward a common goal. All of the requisite tasks to accomplish this goal need to be assigned to the appropriate organizations, monitoring needs to reveal how well the actions are performed, and all of the efforts need to be orchestrated with some form of command and control. As argued in a previous report, that enterprise-wide effort has not yet fully matured.[15] Our goal in this report is not to rehash this aspect of the issue. But it is important to note that the more mature the coordinated enterprise effort is, the more successful actions at the wing level will be for cyber mission assurance. Absent a coordinated enterprise effort, there are significant limitations to how well the problem can be solved at the wing level.

With that caution, we outline here some of the key organizations that directly or indirectly support wing cyber mission assurance. In Chapter 3, we take up how these interactions and support could be improved. But absent a coordinated enterprise effort, these improvements will be incremental, piecemeal advancements.

Program Management Offices

PMOs provide life-cycle management of weapon systems. PMOs are essentially responsible for a weapon system inside the system boundary. The owning and operating command is responsible for everything outside the system boundary, such as the tactics, techniques, and procedures for employing the weapon system. To the extent that an organization within the Department of the Air Force develops, maintains, and sustains technical knowledge of cybersecurity matters of weapon systems, that knowledge lies in PMOs. To the extent that such knowledge exists outside the Department of the Air Force, it resides in the contractors who designed and manufactured the systems. The PMO is the primary interface between the Department of the Air Force and those contractors.

Cyber Resiliency Office for Weapon Systems

The Cyber Resiliency Office for Weapon Systems (CROWS) was established by the acquisition community in the Department of the Air Force to "bake in" cyber resiliency in new weapon systems and to mitigate critical vulnerabilities in fielded weapon systems. It reached full operational capability in 2017 and, as of 2020, resides in the Air Force Life Cycle Management Center under Air Force Materiel Command. CROWS provides specialized knowledge in weapon system cybersecurity and cyber incident response for weapon systems.[16]

---

[15] See Snyder et al., 2021.

[16] For more information on the role of CROWS, see Cyber Resiliency Steering Group, 2016.

Authorizing Officials

Cyber mission assurance differs from other dimensions of mission assurance in that it heavily involves matters that go beyond operational procedures. It also includes the security of relevant hardware, software, and firmware. Federal law inserts the Chief Information Officer (CIO) in every federal agency, including the Department of the Air Force, into cyber risk acceptance for cybersecurity of these systems. The statutory duties of the CIO for weapon systems are largely executed by Authorizing Officials (AOs), who are the individuals "with the authority and responsibility for accepting risk for an IT system."[17] AOs grant approval to connect between systems and approval to operate of systems. Although policy directs AOs to consider operational risk, in practice, AOs focus on system risk and system risk mitigation. For this reason, and because weapon system AOs reside in Air Force Materiel Command, information flow between the AO and the PMO is stronger than the flow between the AO and the operating units, including wing-level organizations.

16th Air Force

Among other missions, 16th Air Force performs defensive cyberspace operations, performs U.S. Department of Defense (DoD) information network operations, provides Cyberspace Protection Teams (CPTs) under the 67th Cyber Wing, and presents Air Forces Cyber to U.S. Cyber Command. CPTs are teams trained in cyberspace defense operations, with an emphasis on IT and ICS. They can provide support remotely or by deploying to a relevant location. CPTs support national and combatant command needs, but some are service-aligned and can be called upon for cyber incident response in the Department of the Air Force. In that capacity, CPTs can augment wing-level capabilities.[18]

16th Air Force exercises command and control through the 616th Operations Center (616 OC).[19] Although the main thrust of cybersecurity operations in 16th Air Force is networked IT systems and not weapon systems or ICS, it has broad authorities that extend to nearly any system that processes DoD information within the Air Force Information Network.

For weapon systems in the Department of the Air Force, exact bounds of authorities have not been fully resolved. Exactly where cyber-related authorities begin and end for weapon systems is unclear for the operating command, the mission owner, the PMO, the CIO and AO, and 16th Air Force. These ambiguities complicate efforts at the wing level.[20]

---

[17] Air Force Instruction 17-101, 2020, Section 3.3, p. 9. AOs are key players in the National Institute of Standards and Technology Risk Management Framework.

[18] Joint Publication 3-12, 2018.

[19] 16th Air Force, 2020; Joint Publication 3-12, 2018. See also Cohen, 2019.

[20] See Snyder et al., 2021, for further discussion.

A final key entity that supports wing-level organizations is the collective intelligence and counterintelligence communities. Both provide strategic and tactical intelligence information for better situational awareness. They are also capable of assisting in forensics of any suspected malware. Intelligence support typically flows through the Operations Group of a wing. One structural challenge is the strong legal separation of foreign and domestic intelligence collection, and a second is the limited distribution of information collected for use in a criminal investigation. These can limit sharing of information with a wing. Classification levels sometimes also present barriers.

Ongoing Initiatives

To better address cyber mission assurance, the Department of the Air Force has embarked on several initiatives that were ongoing during this research. The two most important for wings are Enterprise IT as a Service (EITaaS) and MDTs, both part of the larger Cyber Squadron Initiative. We describe just the broad contours of these initiatives because they are evolving, and this research is intended to help shape future iterations of these policies rather than be bound by them.

Enterprise Information Technology as a Service

In the Mission Support Group of a host wing, or in the Air Base Wing, the Communications Squadron has been the "wing focal point for all cyberspace operations and planning" and has interfaced directly with external organizations regarding cyberspace matters for the wing.[21] In an aim to improve network services, the EITaaS initiative is transitioning many of the IT services traditionally provided by the Communications Squadrons to commercial providers. What were formerly Communications Squadrons are becoming Cyber Squadrons, with the intent of ultimately aligning the new Cyber Squadrons under the Operations Group. This shift of responsibilities frees U.S. Air and Space Force manpower positions, which are reallocated to a more direct warfighting role in a Cyber Squadron and in newly formed MDTs.[22]

Mission Defense Teams

The Department of the Air Force created MDTs to improve what we call *cyber mission assurance at the wing level*. The ultimate goal of MDTs is to assist wing-level commanders in maintaining mission assurance in the face of adversary cyber operations, both at home station and when elements of the wing deploy. In practice, MDTs currently focus on defending specific

---

[21] Air Force Instruction 38-101, 2019, p. 113.

[22] See, for example, Corey and Strobel, 2018. For details of the new Cyber Squadrons, see Headquarters United States Air Force, 2020.

weapon systems via defensive cyber operations at the tactical level.[23] Unlike CPTs, the Department of the Air Force retains control of MDTs. They are not to be assigned to combatant commanders, although, to support deployment, some MDTs are forming UTCs, which would deploy to a geographic combatant commander when tasked to support a deployed weapon system.

Because many wings operate similar or identical weapon systems, lead MDTs are selected to develop tactics, techniques, and procedures for the defense of specific weapon systems. These MDTs have served as pathfinders for the initiative and are more mature than other MDTs that were established later.

MDTs are expected to operate at the tactical level and receive assistance from outside the wing. This reachback to other organizations for support is meant to be assisted by Cyberspace Defense Correlation Cells (CDCCs). CDCCs also facilitate situational awareness for the MDTs. CDCCs, still in the early stages in 2020, are intended to coordinate efforts among MDTs defending similar weapon systems. As command and control are needed for MDT response to cyber incidents, the Cyberspace Defense Operations Center (CDOC) within the 616 OC will take on those responsibilities.[24]

Standardized training and tools were still being developed in 2020 for MDTs. One part of the training that we will discuss more fully in Chapters 2 and 3 helps wings map their mission space to reveal the cyber attack surface and critical systems. That training is called *Functional Mission Analysis-Cyber* (FMA-C) and is offered by the Air University. The key, common tool for MDTs is the Cyberspace Vulnerability Assessment/Hunter (CVA/H) weapon system.[25] CVA/H is a weapon system in its own right, under PMO oversight.

The common metaphor in the Department of the Air Force is that MDTs are like beat cops. They monitor weapon systems day to day at the tactical edge and handle simple problems. When the situation gets more complicated, they call in CPTs, which are like SWAT teams; they handle more-complicated cyber incidents.[26]

Research Approach

We address two central questions:

- What can wing-level organizations do for mission assurance in the face of adversary cyber operations, both at home station and when deployed?

---

[23] A *weapon system* is "a combination of one or more weapons with all related equipment, materials, services, personnel, and means of delivery and deployment (if applicable) required for self-sufficiency" (Joint Publication 3-0, January 17, 2017, Incorporating Change 1, October 22, 2018, p. GL-17).

[24] For more on CDCCs and CDOC, see Air Combat Command, 2018.

[25] See Carter, 2019.

[26] For more on MDTs, see Air Combat Command, 2020; and Headquarters United States Air Force, 2020.

- What should other organizations do, and what organizational mechanisms are needed, to support wing-level organizations?

We took several approaches to these questions. One major thrust was a review of relevant literature on managing security issues within organizations and the specific challenges posed by nation-state-level cyber operations. We supplemented this research with examination of the organizational design literature in sociology, focusing on best ways to apportion roles and responsibilities.

To understand the challenges faced by wings and to solicit ideas for improvement, we conducted a wide range of exploratory interviews, using an open-ended but semi-structured format, of wing-level commanders, Communications Squadron commanders, MDT members, and other cyber stakeholders at the wing level. We performed a structured analysis of numerous interviews that form underlying data for many of our findings. The methodology used is described in Chapter 2 and the appendix. We also interviewed a number of others within the Department of the Air Force and the intelligence community. These included Headquarters, Department of the Air Force; various units in Air Combat Command (ACC); Air Force Materiel Command; Air Force Space Command (now the U.S. Space Force); the National Air and Space Intelligence Center (NASIC); U.S. Army Cyber Command; U.S. Army Program Executive Officer for Simulation, Training, and Instrumentation; and the Department of the Army's Management Office-Cyber.

In the next chapter, we summarize key observations from the interviews, focusing on wing commanders, MDT members, and members of the intelligence community. Chapter 3 presents concrete, actionable ways to improve wing-level cyber mission assurance. Chapter 4 recommends necessary management steps to enable the recommendations in Chapter 3.

2. Empirical Observations at the Wing Level

> Leaders and their staffs need to be "cyber fluent" so they can fully understand the cybersecurity implications of their decisions and are positioned to identify opportunities to leverage the cyberspace domain to gain strategic, operational, and tactical advantages.
>
> — DoD[27]

As this research was being conducted, the initiatives to create Cyber Squadrons; provide EITaaS; and create, train, and develop MDTs were underway. It was important to understand the perspectives on these initiatives at the wing level from early participants in the implementation of these initiatives and from the pathfinders. We also wanted to understand the challenges that each wing is facing, to understand early successes that were enabling implementation of these initiatives, and to glean any insights from wing-level personnel for ideas for improvement of cyber mission assurance.

To these ends, we conducted a number of interviews of wing-level personnel in a variety of wing-level organizations. This chapter summarizes what we learned from these interviews, as well as from selected additional interviews outside wings. The purpose of this chapter is twofold. First, it provides the basis for the findings in the next chapter. Several of the findings are further elaborated in the next chapter to motivate specific recommendations. Second, this chapter provides a record of the challenges in implementing the initiatives at the wing level during 2020. Because we limit our recommendations in the next chapter to ones that are relatively easily implemented and that are most likely to have positive impact, we do not seek remedies for all of the findings. Yet systematically describing the results of the interviews independently of the recommendations will help manage these initiatives as they evolve over the years.

We interviewed personnel to capture how they perceive the problem of cyber mission assurance and the current initiatives within their units, which are called their *mental models* of these issues.[28] These mental models reveal a number of issues to rectify, including both issues that are not adequately addressed by the current initiatives and policies and areas in which the initiatives and policies are falling short of their intent.

That is not to say that we take all of the statements of the participants as fact. When interviewing experts, it is important to understand what they are experts in.[29] Some may hold parochial views of the problem; be misinformed or underinformed about details; or fall victim to

---

[27] DoD, 2018, p. 5.

[28] See Morgan et al., 2002.

[29] Morgan, 2014.

cognitive heuristics, or mental shortcuts, that lead to misrepresentation of the problem.[30] But those observations provide useful insights. They reveal gaps in policy and disconnects between how leaders in organizations perceive problems and how working-level personnel perceive them. The sum of the observations in this chapter form the foundation of our findings.

Methods

Wing-Level Interviews

We conducted 37 interviews with wing commanders, MDT members, and other relevant organizational elements from 16 wings with a diverse cross-section of wing-level characteristics. Table 2.1 provides a list of the wings we interviewed, characterized by Major Command (MAJCOM) and mission area, as well as by whether the MDT was designated as a lead MDT, whether the principal weapon system is an aircraft or a networked system, whether the weapon system operators deploy, and whether the wing is the host unit for its base.[31] Interviews were open-ended and semi-structured to cover the following topics: roles and responsibilities; tasks, activities, and decisionmaking; challenges and potential improvements; information flow and organizational interactions and support; cyber situational awareness; and cyber incident processes, among others. Notes were captured for all interviews; these notes formed the data source for our qualitative analysis.

We performed thematic coding and analysis on 27 of the 37 interviews—those in which we spoke with wing commanders and MDT members. This method for qualitative analysis identifies emerging themes from the interviews. We first tagged excerpts of text from the interview notes according to themes. We then used these tags to transform qualitative text into quantitative data for structured analysis.[32] We use analysis of these data themes, for example, to systematically explore the patterns and interrelationships of cyber mission assurance concepts discussed by interviewees and to make comparisons of these results across interviews.

It is important to note that this analysis is not meant to provide results that are generalizable to the entire population being studied. Rather, it is intended to provide the range of possible concepts the population is considering and to explore potential differences among populations. Similarly, our intent in using the approach was to explore the range of beliefs, knowledge, and current conditions affecting wings' ability to ensure cyber mission assurance. Any results from qualitative analyses should not be seen as generalizable to all Department of the Air Force operational wings or different types of wings.

---

[30] Kahneman, 2011.

[31] A number of the captured characteristics are not entirely binary (e.g., a single wing may perform missions that both do and do not deploy aircraft). Here, we represent the characteristic for each wing that is most dominant.

[32] Braun and Clarke, 2006.

We developed our interview themes and resulting analyses to answer the following research questions from the perspectives of these wings:

- What are the primary challenges wings face for ensuring cyber mission assurance?
- What are the enablers to wings as they work to ensure cyber mission assurance?
- What level of cyber-related interactions and support do wing commanders and MDTs have/receive from within the wing and from other organizations?
- What enterprise- and/or wing-level changes could help to improve wings' cyber mission assurance?
- How do perspectives on the above questions differ between wing commanders and MDTs and among wings with different wing-level characteristics (i.e., those shown in Table 2.1)? For the latter, what do these characteristics reveal about the unique aspects of wings to be considered when developing enterprise-wide improvements for cyber mission assurance?

These five research questions were designed to inform answers to the two central research questions presented in Chapter 1. To answer what wing-level and other organizations can do for mission assurance, an understanding of wings' current experiences was needed to properly ground our findings in reality. Because the initiatives were rapidly evolving, we did not focus questioning on the state of the efforts at the time of the interview. Rather, we focused on those aspects of the implementation that participants found enabled or challenged their best efforts to carry out the initiatives' goals.

Table 2.1. Wing-Level Interviews

Wing (or Equivalent)	MAJCOM	Principal Weapon System(s)	Mission Area	Lead MDT	Aircraft	Deploys	Networked System	Host Wing
53rd Wing	ACC	EW/Test	Test, EW	X	X			
55th Wing	ACC	RC/OC/TC/WC-135	ISR	X	X	X	X	X
480th ISR Wing	ACC	DCGS	ISR	X			X	
552nd Air Control Wing	ACC	E-3	C2	X	X	X		
1st Fighter Wing	ACC	F-22	Fighter		X	X		
9th Reconnaissance Wing	ACC	U-2	ISR		X	X		X
601st AOC	ACC	AOC	C2				X	
22nd Air Refueling Wing	AMC	KC-135, KC-46	Air refueling	X	X	X		X
89th Airlift Wing	AMC	VC-25, C-32, C-37, C-40	Executive airlift	X	X			
341st Missile Wing	AFGSC	Minuteman III	ICBM					X
509th Bomb Wing	AFGSC	B-2	Bomber		X	X		X

Wing (or Equivalent)	MAJCOM	Principal Weapon System(s)	Mission Area	Lead MDT	Aircraft	Deploys	Networked System	Host Wing
50th Space Wing	USSF	AFSCN	Space	X	X		X	X
460th Space Wing	USSF	SBIRS	Space				X	X
613th AOC	PACAF	AOC	C2	X			X	
603rd AOC	USAFE	AOC	C2				X	
482nd Fighter Wing	AFRC	F-16	Fighter	X	X	X		X

NOTES: AFGSC = Air Force Global Strike Command; AFRC = Air Force Reserve Command; AFSCN = Air Force Space Control Network; AMC = Air Mobility Command; C2 = command and control; DCGS = Distributed Common Ground System; EW = electronic warfare; PACAF = Pacific Air Forces; SBIRS = Space-Based Infrared System; USAFE = U.S. Air Forces in Europe; USSF = U.S. Space Force. The MDT assigned to the 1 Fighter Wing is organized under the 633 Air Base Wing. The 509 Bomb Wing had not stood up its MDT. An X in a column indicates that the wing has that characteristic; the absence of an X indicates that it does not.

The appendix presents comprehensive discussion of our wing sample selection; interviewing methods; and analysis procedures, including methodological foundations, interview protocols, and the full set of themes.

Intelligence Community Interviews

In addition to the interviews at the wing level, we wanted to understand how the intelligence community, particularly Department of the Air Force intelligence units and components, are supporting the cyber resiliency mission, both from a day-to-day perspective of understanding the cyber threats to a wing's mission and in supporting cyber incident response. The following are the main questions we sought to address through these interviews:

1. What is the current process for providing cyberspace intelligence support to wings and MDTs?
2. What kind of intelligence support is available to MDTs?
3. Which units are responsible for providing intelligence support, and to what extent are they able to meet the demand?
4. Where and how does the current process fall short?

To answer these questions, we conducted interviews with a variety of organizations responsible for collecting, analyzing, and disseminating intelligence to the operational force. We did not interview every unit that provides intelligence to wings, but we were able to interview the primary units responsible for providing intelligence related to cybersecurity. We then used snowball sampling to identify additional units and organizations to interview.[33] In some cases (e.g., 616 OC Intelligence Surveillance Reconnaissance Division [ISRD]), a unit or organization

---

[33] *Snowball sampling* asks initial interviewees for recommendations of others to interview, and then asks the next set of interviewees for further recommendations, and so on until research objectives are reached. See Naderifar, Goli, and Ghaljaie, 2017.

is not directly responsible for collecting and analyzing intelligence but is involved in the dissemination of intelligence.

We also interviewed representatives from several organizations that engage in sharing information related to cybersecurity issues, such as vulnerability analyses of Blue systems. These vulnerability analyses are not an intelligence mission but are a critical part of understanding the cyber terrain. Given the smaller number of intelligence organizations and units and the diversity of missions, roles, and responsibilities they cover, we did not conduct thematic coding to the degree that we did for the wings and MDTs. Instead, we organized observations and comments across main themes: organizational responsibilities, resources and staffing, tools, intelligence support to mission analysis, and intelligence support to incident response. Table 2.2 lists the organizations that we interviewed to address the intelligence support questions.

Table 2.2. Intelligence Organizations Interviews

Organization	Mission
NASIC/Air and Cyberspace Intelligence Group	Engineering-level scientific and technical intelligence on cyber threats
70th ISR Wing	Global ISR in air, space, and cyberspace
70th ISR Wing/659th ISR Group/ 35th Intelligence Squadron	Intelligence support to national and Department of the Air Force cybersecurity, cyberspace exploitation, operation of the Cyberspace Threat Intelligence Center
70th ISR Wing/659th ISR Group/ 7th Intelligence Squadron	Computer network analysis, development, collection, and defense, by, with, and through partner agency (National Security Agency); cyber vulnerability and threat interdiction
616th OC/ISRD	Strategy and plans for cyberspace operations
616th OC/ISRD	Integration of global ISR in tasking orders and targeted ISR support
ACC CDCC	Interface between Air Force Cyber forces and MDTs, correlation of cyber analysis and incidents
ACC A2/3/6K	Intelligence integration support to MDTs
CROWS	Cyber resiliency of weapon systems
AFMC/A2	Intelligence support to PMOs
AFMC/21 IS	ICS cyber vulnerabilities, threats

Wing-Level Empirical Observations

Common Themes

We identified 11 broad topical themes that dominated our discussions with wing commanders and MDTs. Wings discussed these themes in terms of three associations to cyber mission assurance: as a *challenge* to or an *enabler* of achieving it or in terms of ideas that the Department of the Air Force or wings could implement to *improve* wing-level cyber mission

15

assurance. For each of these associations, we identified patterns within the 11 themes across the 27 interviews.

Overall, wings tended to discuss themes in terms of their challenge more often than in terms of their enabling ability or as improvement ideas, suggesting that, in the current environment, wings have found more problems than solutions related to cyber mission assurance. Still, many of the same themes were discussed in terms of all three associations. Given the broad nature of these themes, our analysis found that wings are finding success with some aspects of these concepts, while other aspects still frustrate them. Furthermore, as we will discuss later, the unique characteristics of each wing may lead to different experiences.

Table 2.3 provides the prominent themes discussed as a challenge, enabler, or improvement idea for 27 wing commander and MDT interviews. Results for wing commanders and MDTs are shown both separately and together. Each theme is briefly discussed in the next section. Note that a single theme can be both a challenge and an enabler—one existing aspect facilitates cyber mission assurance, but the lack of another aspect hinders it.

Challenges

Personnel. Concepts related to wing-level personnel were the most prominent challenge discussed by wings. Personnel challenges covered a range of topics, including the number of individuals needed, their level of relevant expertise and skills, and attrition of personnel.

Training. The needed training of these personnel, once they are stationed at the wing, was commonly mentioned as a challenge. Training challenges included those to specific types of training, especially FMA-C, but also covered the length of training or the wait to receive training.

Authorities and policy. Discussions related to challenges with individual authorities and Department of the Air Force policy were another prominent theme of both wing commanders and MDTs. The challenges covered the lack of guidance on the MDT's role, as well as authorities that individuals needed to make cyber mission assurance decisions.

Equipment and tools. The appropriate tools for the MDT were another prominent area of discussion during interviews, mostly in terms of the challenges wings were currently experiencing. Many of the challenges discussed focused on the CVA/H toolkit.

Relationships and information flow. Challenges related to how information flows and the nature of interactions or relationships between individuals or organizations were prominent topics of concern, but only in MDT interviews.

Resources and funding. The resources required to organize, train, and equip the MDT were commonly discussed by wing commanders. With the Cyber Squadron Initiative Program Action Directive having not yet been signed when these interviews were conducted, wing commanders were quick to point out that their MDT was "taken out of hide."

Cyber culture and priority. Culture related to cyber mission assurance and how this relates to priority placed on organizing, training, and equipping wings were discussed both explicitly and implicitly throughout the wing commander interviews. Cyber priority challenges focused mainly on priorities above the wing.

Access. The ability to access needed systems and information was another recurring challenge to cyber mission assurance wings discussed by MDTs. Prominent topics mentioned included issues with authority to connect to the weapon system and MDT members obtaining necessary clearances.

Infrastructure and facilities. Challenges about the support provided for cyber mission assurance by infrastructure and facilities were mentioned in a few interviews. Although this was not a particularly prominent discussion point for wing commanders or MDTs, those who did mention it often focused on the need for facilities used by MDTs, such as sensitive compartmented information facilities.

Enablers

Training. Not all existing trainings were problematic for wings. MDTs consistently praised within-wing mission-level training and some initial qualification training, resulting in this theme being another prominent enabler.

Relationships and information flow. While MDTs found relationships within and above the wing to be challenging, they also stated that the relationships they had established were extremely valuable. Accordingly, this theme was the most prominent enabler discussed by both wing commanders and MDTs.

Resources and funding. When wing commanders properly resourced their MDTs, discussions about funding were stated to be an enabling function of cyber mission assurance.

Processes and procedures. Cyber-related processes and procedures, such as FMA-C and incident response procedures, were also discussed as a prominent enabler during MDT interviews.

Improvements

Personnel. The most prominent improvement mentioned by wings was personnel, and personnel discussions often focused on creating a career field to allow continuity of institutional knowledge for MDT members.

Training. The challenges with FMA-C training brought about a number of ideas by MDTs for how it could be improved, resulting in this theme being prominently discussed in terms of improvements.

Relationships and information flow. Improving the level of integration of activities and communication was an oft-discussed concept under this theme.

Organizational structure. The relationship of organizational structure, reporting relationships, and roles and responsibilities, both within and outside the wing, to a wing's cyber mission assurance was raised mostly when wing commanders and MDTs discussed their ideas of improvement.

As Table 2.3 shows, wing commanders and MDTs had somewhat different mental models of the challenges, enablers, and improvements to wing-level cyber mission assurance. For example, the results suggest that while both found that within- and outside-wing relationships are prominent enablers to cyber mission assurance, resources and funding was prominently discussed as an enabler only by wing commanders. Of course, some of these disparities reflect tenure, professional values, and different positions within the wing. However, it is also possible that these differences suggest a disconnect between the needs and experiences of MDTs and those perceived by the wing commander.

Table 2.3. Prominent Themes Discussed in Interviews

Theme	Challenge			Enabler			Improvement		
	All	CC	MDT	All	CC	MDT	All	CC	MDT
Personnel	■	■	▒		▒		■	■	■
Training	▒	▒	▒	▒	▒		▒	▒	▒
Authorities	▒	▒	▒						▒
Equipment	▒	▒	▒						
Relationships	▒		■	■	■	■	▒		▒
Resources	▒	▒		▒	▒				
Culture		▒				▒			
Access	▒		▒						
Processes				▒		▒			
Organization							▒	▒	▒

NOTES: CC = wing commander. Shaded boxes denote those theme and comment types for which discussions were prominent, separated by the stakeholder interviewed (i.e., wing commander, MDT, and all interviews). Dark blue boxes denote the most prominent theme for each comment type and stakeholder interviewed.

Training and relationships were highlighted as challenging areas by MDTs. And, although our analysis suggests that wing commanders are aware of those challenges, they may not understand the full magnitude of them. Similarly, processes and procedures, specifically those for FMA-C, were highlighted as enablers by MDTs, yet this theme was not commonly discussed by wing commanders.

When wings with different wing-level characteristics (e.g., host versus tenant wings) are compared, the patterns discussed generally hold for challenge themes but vary somewhat for discussions about enablers. This observation suggests that, at a high level, the challenges that wings face do not seem related to their unique characteristics,[34] but the enablers they identify may be more related. Stated another way, most wings face broadly similar challenges, but the

---

[34] However, patterns within each theme do differ for some wing-level characteristics, suggesting that some wings have more difficulties in some areas than other wings. This observation will be discussed later in the chapter.

most-effective means of facilitating their cyber mission assurance may vary based on their unique circumstances. However, relationships and information flow were the most prominent enabling theme across almost all wing-level characteristics tracked,[35] suggesting that communication and within- and outside-wing networks are helpful to ensure cyber mission assurance regardless of the unique aspects of a wing.

The 11 themes discussed thus far are necessarily broad. Our review of interview discussions within many of these themes elucidated a more nuanced narrative. We therefore developed a number of subthemes to explore each theme at a more detailed level. In the following sections, we highlight interview insights from analysis of these subthemes.

Resources

Although wing commanders we interviewed varied in their prioritization of cyber mission assurance activities to ensure the wing's mission, many noted that resourcing these activities requires them to make difficult trade-offs with other wing priorities. For example, one wing commander discussed using squadron innovation funds to send MDT members to courses and conferences, funds that could have been used elsewhere. Furthermore, the most effective way to allocate cyber mission assurance resources was not clear to many wing commanders, as they had not received above-the-wing guidance on wing-level strategies to ensure cyber mission assurance. All recognized, however, that cyber mission assurance requires knowledgeable personnel who are trained and equipped with the proper tools.

Personnel concerns expressed by both wing commanders and MDTs were largely related to the inability to properly man the MDT, both in terms of needed quantities (e.g., billets) and expertise or skills. There was less frequent discussion about losing trained MDT members to attrition or deployments. While manning issues were mainly attributed to resource constraints, wings discussed that the lack of skilled MDT members was often a result of lengthy training timelines in which fully trained MDT members often had very little time to use their newly acquired skill set before receiving permanent change-of-station orders. Indeed, many of the discussed ideas for personnel improvements focused on sustaining this acquired expertise through MDT career field development.

Challenges discussed surrounding MDT training and tools were often intrinsically linked. For example, given that many of the wings we interviewed do not operate internet protocol–based weapon systems, many found the internet protocol–based CVA/H toolkit to be insufficient to monitor weapon system activity. Some further described not frequently receiving technical updates to the toolkit or needing to rebuild the toolkit each time they did because of system crashes. To work around many of these issues, wings developed in-house training to supplement the initial qualification training they had received. It is this in-house training, along with wing

---

[35] One exception was for wings that are not the host unit on their base. In this case, interviewees most prominently discussed training as an enabler.

commanders using squadron innovation funds for training, that wings stated as an enabler to their cyber mission assurance.

However, the most commonly discussed training subject matter with which wings expressed discontent was that for FMA-C. From a conceptual standpoint, MDT members stated that FMA-C was useful in helping them develop the necessary operational mindset. FMA-C training, though, did not provide them with useful case-study applications, making the training nearly useless for cyber terrain mapping of their specific mission. The classroom examples were too simple relative to the complexity of their wing's mission. For example, one MDT stated that the example used in the course was for elevators opening and closing, noting that their wing has "more than elevators." Still, wings' improvement ideas for FMA-C expanded well beyond more-applicable, case-based training. Some wings discussed having field teams of outside subject-matter experts perform, guide, or provide quality control of their cyber terrain mapping.

Across the various wing-level characteristics we tracked, we found that training difficulties were discussed more by wings with lead MDTs, who in turn were also more likely to lament the challenges of MDT member expertise. Possibly, these more mature wings are more aware of the challenges they are facing. Wings with lead MDTs were also more likely to discuss that proper staffing levels facilitated their successes. Possibly, because wings without lead MDTs might have lower MDT manning, they have not evolved to a place of questioning the training or expertise of those personnel. Wings with networked systems found the MDT training to enable their cyber mission assurance more than wings with non-IT weapon systems, as this curriculum is more focused around internet protocol-based systems. Surprisingly, we also found that these wings were also more likely to discuss challenges with the CVA/H toolkit because of issues with technical updates. Still, a number of wings mentioned having no challenges with CVA/H because, as they explained, they did not have the authority to connect it to their weapon systems. This and other authority-related challenges are discussed in the next section.

Authorities

Our interviews were performed in the first half of 2020, at a time in which the Program Action Directive for the Cyber Squadron Initiative had not yet been signed.[36] Possibly as a result, many discussions focused on the lack of policy and guidance provided to wings on the roles, responsibilities, and authorities of wings to ensure cyber mission assurance. Indeed, a majority of discussions related to challenges with policy and individual authorities centered around the lack of guidance for MDTs. For example, one wing commander who admitted to not having extensive cyber expertise bluntly stated his concern about MDTs, saying, "How do I know they are effective?" Commanders often remarked that they had little to no above-wing guidance on goals for cyber mission assurance, making it difficult to measure how well their wing was ensuring cyber mission assurance. Wing commanders, in turn, generally did not give specific guidance, or

---

[36] Headquarters United States Air Force (HQ USAF), 2020, Not available to the general public.

a commander's intent, to their MDTs. The lack of guidance to MDTs may explain why, when asked about their mission, only two of the MDTs discussed defending the wing's mission. The remaining MDTs were mostly only able to list their activities. Further guidance desired by a number of wings was that for cyber incident response. Some wing commanders, for example, mentioned ill-defined incident thresholds or reporting relationships.

Across the various wing-level characteristics we tracked, challenges related to MDT guidance were more prevalent for wings without lead MDTs, wings that deploy, and those that serve as the host wing at their air base. MDTs who are not the lead are usually less mature and therefore need more guidance. For the latter two wing-level characteristics, it is possible that navigating the MDT's role in these situations (being CHOPed, defending ICS) is more complex, prompting the desire for more guidance.

Additional common challenges with policy and authority that wings discussed were the limitations to their cyber-related individual authorities and jurisdiction. Wings lamented that these limitations were one of the primary reasons for their lack of access to needed information and systems. Across discussions about access, wings relayed concerns from AOs about inexperienced personnel inadvertently causing problems with the weapon system that inhibited the granting of authority to connect. Those wings with this challenge stated that such lack of access inhibited their abilities to develop mission-specific training, perform comprehensive system defense, and gain much-needed experience. In wings with access to their weapon systems, however, MDTs highlighted this as a major enabler to defending the weapon system. Other than authority-to-connect challenges, issues with obtaining proper clearances were prominently discussed. Wings operating networked systems were more likely to discuss these issues, as these systems often required accessing information at higher classification levels.

To ameliorate the challenges with authorities and access, some wings stated that roles and responsibilities within and outside the wing should be better defined. Ideas for improvement in this area ranged from clearly defining the role of members of the MDT as cyber operators to specifying the reporting relationships and roles with a wing's parent MAJCOM, the Numbered Air Force, CDCC, and CPTs. Wings discussed how, without these roles and responsibilities clearly defined, their current strategy is highly reliant on building good relationships with outside stakeholders. These relationships are further discussed in the next section.

Relationships

Given the limited authorities and guidance, as well as the uncertain nature of the wing's role in ensuring cyber mission assurance, wings that we interviewed discussed the need to develop their own mechanisms for communication and information flow. These improvisations have been met with mixed results for relationships both outside the wing and between the MDT and other wing-level organizational elements.[37] Indeed, the results shown in Table 2.3 suggest that although relationships within and outside the wing are challenging, wings believe that the good relationships they have established are the largest enabler to ensuring cyber mission assurance. MDTs expressed the view that they build these enabling relationships by proving their value to organizations both within and outside the wing. However, many lamented that they still are siloed from other organizational elements in the wing. They discussed such silos existing for a number of reasons, including that many MDTs are organized under a wing's Communications Squadron, as opposed to the operations squadron, and, relatedly, that Air Force culture treats cyber personnel similar to IT personnel rather than as "cyber operators," as many MDT interviewees view themselves.

Relatedly, a number of wings suggested ideas for improvement to the within-wing organizational structure that they believed would enable integration of MDT activities with operations, maintenance, and especially intelligence. MDTs often complained that actionable cyberspace intelligence support was a challenge because they did not have a point of contact, because the wing lacked personnel that could translate intelligence to cyber-specific needs (and vice versa), or because MDT members lacked the requisite clearances to obtain such information. Such integration of cyber mission assurance activities, both within the wing and between the wing and other organizational elements, was discussed as an enabler to wings in a number of enabling-relationship comments. In host wing discussions, a prominent issue raised was the need for information flow with all tenant units to facilitate and integrate plans and response. Tenant units need to share information only with the host unit.

Another prominent area of discussion related to relationships and information flow was the sharing of best practices and lessons among MDTs. The more-mature wings with lead MDTs discussed playing more of a mentoring role to the less-mature MDTs. However, the relationships were mostly developed in an informal way, such as through conference interactions. Possibly for this reason, wings without lead MDTs were more likely to discuss relationships and information flow as a challenge. Although many MDTs stated that they would not want such interactions to be prescribed, a number of their improvement ideas in this area did highlight the development of more-formal information-sharing platforms.

Much of the remaining discussion about relationships focused on the nature and extent of wing interactions with other organizational elements or individuals, as described in the next two sections.

---

[37] We defined *relationships* broadly for our analyses, including those ranging from formal to informal and unidirectional to balanced, as well as those that consisted of anything from a one-time interaction to frequent interactions.

Challenges and Enablers

Our qualitative analysis of challenge and enabler discussions, in which specific organizational elements or individuals are mentioned, may provide some insights into those stakeholders for which wings determined there were the most relationship challenges or successes. The MDT training entities, as well as the MDT, were prominently discussed in terms of both challenges and enablers to cyber mission assurance. As previously stated, being designated to both categories may not be inaccurate. Indeed, MDTs routinely discussed having both positive and negative experiences with different types of training. And wing commanders generally were very impressed by what their MDTs had accomplished but admitted that many challenges remained.

A number of MDTs routinely discussed how, in turn, wing commanders enabled their ability to better perform their role. That is, MDTs stated that when a wing commander prioritized cyber mission assurance and provided MDTs with buy-in, the wing's cyber culture provided them with much-needed leverage and authorities both within and outside the wing. In contrast, many wing commanders described how the lack of authoritative guidance and resources from their parent MAJCOM and the Department of the Air Force signaled a lack of cyber culture above the wing. In their view, this challenged their ability to ensure their wing's cyber mission assurance by undermining wing commanders' leverage, authority, and autonomy in the cyber domain.

Although the 16th Air Force was commonly described as supporting MDTs by providing relevant threat information, within-wing intelligence information was often lacking for MDTs because of a lack of cyber-specific intelligence, as previously discussed. Many of the wings for which EITaaS had been established also had discussed the challenges of negotiating the currently ill-defined roles in EITaaS contracts.

Operators and the CVA/H PMO were discussed more in terms of playing an enabling role for wings employing a primary weapon system that was not an aircraft, was networked, and does not deploy, suggesting that information flow and relationships with these stakeholders may be more streamlined than for wings with aircraft. Those wings with aircraft, along with those that deploy and those within ACC, also were more likely to mention their relationship with other wings when discussing enablers to their cyber mission assurance, suggesting that the informal networks of these types of wings may be more mature.

Networks Within and Outside the Wing

To understand the existing networks of organizational interrelationships regarding cyber mission assurance, we asked every wing commander and MDT to discuss their current interactions related to cyber mission assurance within and outside the wing. From this open-ended question, we identified eight within-wing and 13 outside-wing organizational elements or individuals discussed by at least two wing commanders or MDTs. Although the nature of these interactions varied (e.g., whether they were challenging or enabling, as described in previous sections), as did their type (e.g., ad hoc discussions, conferences, reporting relationships), the presence or absence of an interaction suggests whether an organizational element is a part of a wing's cyber mission assurance network.

Using the presence of an interaction to define a wing commander's or MDT's network, we found that the size of the networks of these wings varied—a few wing commanders discussed having none or only one of these interactions, while other MDTs discussed as many as 13 existing interactions. On average, MDTs' and wing commanders' interactions were generally limited. Every MDT discussed at least one interaction outside the wing, but two of these MDTs did not provide a single within-wing interaction. Of the eight within-wing and 13 outside-wing organizational elements or individuals we tracked, MDTs averaged fewer than three existing within-wing and fewer than five existing outside-wing interactions. Seventy-three percent of wing commanders mentioned outside-wing interactions related to cyber mission assurance, and just over half discussed any within-wing interactions. Wing commanders averaged less than one within-wing interaction and a little more than two outside-wing interactions in terms of cyber mission assurance.

Figures 2.1 and 2.2 provide the percentage of wing commander and MDT interviews, respectively, in which their interactions with specific organizational elements or individuals were mentioned. The figures show that not only the size but also the composition of these wings' networks varied. Overall, about half of MDTs and wing commanders had interactions with one another (50 percent of MDTs and 40 percent of wing commanders mentioned the interaction). This interaction often was discussed in the form of a one-time briefing, but a few did describe more-frequent interactions. Within-wing intelligence was the most discussed interaction for MDTs, but only one MDT mentioned having an interaction with intelligence organizations outside the wing. The latter is not surprising, as a number of MDTs admitted that they did not know whom to contact outside the wing for intelligence. Other outside-wing interactions mentioned by zero or one MDT included Headquarters, Air Force; AOs; and other DoD services. Most MDTs discussed having interactions with other MDTs; these interactions ranged from ad hoc conversations at conferences to information-sharing between lead and non-lead MDTs. About half of the MDTs we interviewed also had interactions with the PMO of their weapon system, other wings, and the CDCC, while about half of the wing commanders discussed interactions with other wings and their parent MAJCOM.

Figure 2.1. Percentage of Wing Commander Interviews Mentioning an Interaction

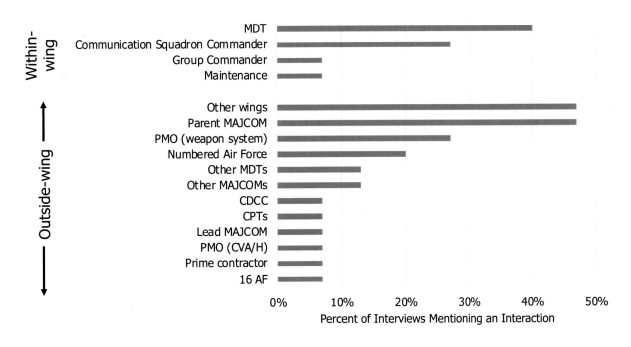

Figure 2.2. Percentage of MDT Interviews Mentioning an Interaction

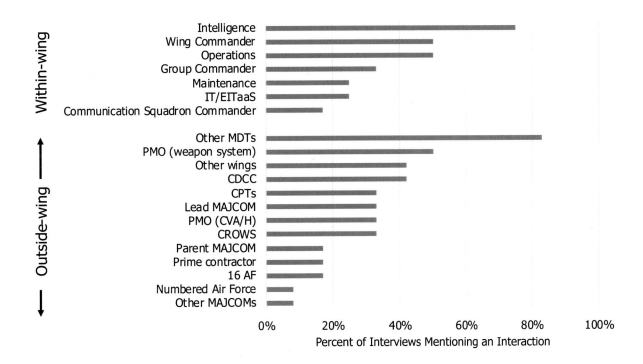

Summary of Wing-Level Interviews

The Cyber Squadron Initiative and the development of MDTs are relatively new Department of the Air Force initiatives. Wings are still in the very early phases of learning the most effective

25

way to ensure their cyber mission assurance. It is normal for wings to be experiencing challenges and learning lessons. Many are beginning to innovate and build strong relationships with relevant organizations to achieve some early successes. However, wings are currently struggling to understand what is required to ensure cyber mission assurance and how to effectively use MDTs.

Much of what wings require to improve further is out of their control. Many of the wing commanders we interviewed attributed these obstacles to an underlying lack of enterprise-level cyber prioritization and guidance. Without guidance, navigating the effective implementation of wing cyber mission assurance requires a "champion," which heavily depends on personalities, expertise, and networks. This model, the wing commanders stated, was an unsustainable one. Furthermore, these wing commanders told us that the lack of cyber mission assurance as a priority in the Department of the Air Force acts as an adverse signal, decreasing their leverage, authority, and autonomy. Most wing commanders and MDTs said that they wanted non-prescriptive guidance to define cyber mission assurance objectives, authorities, roles and responsibilities, career management, and relationships with external organizations.

Since these interviews were conducted, the Program Action Directive for the Cyber Squadron Initiative has been signed. Perhaps this event will provide the shift in priority for cyber mission assurance for which the wing commanders we interviewed were hoping. However, many challenges will remain. Changing cyber culture within and outside the wing will take time, effort, and buy-in from Department of the Air Force senior leaders.[38] As one wing commander explained to us, hopefully this can occur before being triggered by a catastrophic event. Referring to the need for a shift in cyber priority, the wing commander left us with an ominous hypothetical: "What's it going to take [for this change to occur]?"

Intelligence Community Observations

Before diving into our observations from interviews with members of the intelligence community providing support to MDTs, we briefly review the intelligence process. This entire process needs to be considered when evaluating how the Department of the Air Force cyberspace intelligence organizations support MDTs.

The process for requesting intelligence has many steps. According to joint doctrine, the intelligence process begins with conducting joint intelligence preparation of the environment to understand the threats to Blue forces. Drawing on this intelligence preparation, the commander develops commander's critical information requirements (CCIRs). These motivate *prioritized information requirements* (PIRs), which are information requirements concerning the Red threat, and *friendly force information requirements*, which are information requirements on Blue forces. From the PIRs, essential elements of information are developed that inform the development of

---

[38] For a detailed discussion, see Snyder et al., 2021.

requests for information (RFIs). RFIs become intelligence collection and production requirements. Finished intelligence is then disseminated to appropriate parties.[39]

Our interviews with intelligence organizations revealed some consistent themes that affect several steps in the intelligence process: defining what *cyberspace intelligence* encompasses; ensuring that organizational responsibilities are clearly articulated; staffing and training personnel for intelligence support to cyber resiliency missions; implementing a consistent and clear approach to providing intelligence support both in steady state and during incident response; and providing the tools and products for intelligence support.

Defining Cyberspace Intelligence

Defining intelligence requirements is a critical step in the first part of the intelligence process. However, in our interviews, several intelligence units and organizations noted that the concept of *cyberspace intelligence* itself is not well understood or agreed upon, which can lead to mismatches in expectations between requesting organizations, such as MDTs or program offices, and the intelligence organizations seeking to respond to and support these requests. The first issue is that many MDTs are perceived to want highly focused and detailed intelligence on threats to their wing's systems. This manifests in RFIs articulated as, "What is the threat to Link-16?" or "Can Red exploit my weapon system?" This form of question was acknowledged as legitimate but often difficult or even inappropriate for an intelligence organization to address. Answering questions posed in this form requires a detailed understanding of a Blue system and might require exquisite intelligence on an adversary's cyber capabilities and intent that is not known to the intelligence community.

The intelligence organizations we talked to differentiated between their ability to advise on trends in the strategic threat—for example, what an adversary is likely to want to accomplish in and through cyberspace—and the tactical threat—for example, an adversary's discrete capabilities to execute a cyber operation against a particular system. The intelligence organizations also noted that requests that focus on threats to a particular system can often lead to redundant requests and analysis if that system or subsystem is used across multiple units or weapon systems. This issue is supposed to be addressed through the CDOC and CDCC construct, but that construct also makes it incumbent on intelligence organizations to use a standardized system for receiving, processing, and responding to intelligence requests rather than receiving direct queries from tactical units (see the next section).

Finally, intelligence organizations underscored that information about cyberspace vulnerabilities that have been exploited or are subject to exploitation are part of what the commercial world calls *cyber threat intelligence*, but cyber threat intelligence is different from *cyberspace intelligence* as an intelligence agency would develop it. Therefore, commercial cyber

---

[39] Joint Publication 2-0, 2013.

threat intelligence from such firms as CrowdStrike or Anomali are important tools, but supplying this information to wings is not a traditional intelligence function.

Organizational Responsibilities

Intelligence support to MDTs draws on a variety of organizations engaged in the collection, analysis, fusion, and dissemination of intelligence and information. Sources are diverse and include signals intelligence, human intelligence, and open source information. Most MDTs do not have direct insight into who does what in the process, but they or their Operational Support Squadron/Intelligence (OSS/IN) may come across several organizations.

In the initial phase of the MDT initiative, organizational responsibilities were not always clear, and this stemmed in part from a lack of clarity about what intelligence support is needed for cyber mission assurance and what needs to be done to supply it. This led to some tactical units, such as the 35th Intelligence Squadron (IS), stepping up to serve as a primary organization from which MDTs could directly seek intelligence support. With the development of the CDOC and CDCCs, this direct line between MDTs and tactical intelligence units is being severed and replaced with a more formalized process and lines of authority. However, as was made clear in the majority of our interviews, this process was far from complete in 2020 and still faces many challenges, particularly from a lack of resourcing.

Organizations like NASIC generally focus on providing strategic intelligence in terms of adversary intent, trends, and capabilities, including technical cyber capabilities. Other organizations, such as the 7th IS, the 21st IS, and CROWS, focus on cybersecurity threats to weapon systems and installations. These organizations conduct analyses for the program offices responsible for the weapon systems rather than the operational units that operate the platforms. In fact, the few vulnerability assessments of specific Blue systems that are conducted appear to be rarely shared beyond the program office because of classification.

Staffing and Training for Cyberspace Intelligence

Most of the organizations we spoke to said that all-source intelligence analysts were the primary Air Force Specialty Code (AFSC) used for defining intelligence billets supporting the intelligence mission at the wing level. They described personnel who are (or should be) well-trained in intelligence tradecraft but tend to lack training and education on cyberspace concepts and technology. The availability of cyberspace training opportunities is a challenge. The Cyber Intelligence Functional Training Unit at Hurlburt Field temporarily suspended courses and was undergoing a revision to shorten a course and make it more relevant to analysts going into cyberspace intelligence roles at the time of writing. The cyber courses that are referred to as Cyber 100, 200, and 300 are not always available to personnel when needed. Several interviewees noted that combining a deep intelligence background with a computer science or computer engineering degree is unusual but would ultimately be most useful to possess. The question remains whether the Department of the Air Force can scale this training and staffing to meet the need.

In terms of staffing intelligence units, several organizations noted that they expected that the growth in MDTs would place significant demands on the intelligence community as a whole in the coming years. They also expressed that it was not entirely clear that intelligence units and organizations would have the staffing (or systems and processes) to support this increase in demand. As one unit noted, it only had one-quarter of its authorized billets filled, and the authorization predated several of the recent initiatives, including the formal Program Action Directive establishing the MDTs. Within the CDCCs, intelligence billets were not filled as of August 2020, and the current staff lack the expertise and training to provide the requisite support to MDTs. Across the board, units stated that there are far fewer resources available than can satisfy the requirements; even with anticipated future bodies, the perception from the intelligence support perspective is that the "MDT initiative is woefully under-resourced and under-trained."

Processes for Intelligence Support to Cyber Resiliency

Intelligence support to MDTs can be thought of in two forms: intelligence support for the daily activities that MDTs undertake to understand and defend their wing's cyber terrain, and the support required to assist in cyber incident response. A typical intelligence support cycle should start with the MDT working with the wing's OSS/IN to develop and refine the intelligence needs. The OSS/IN may be able to provide the support that the MDT needs through local resources or by "pulling" finished intelligence from the Secure Internet Protocol Router Network (SIPRNet) or higher systems (if available at the installation). This requires, of course, that a wing has an OSS/IN and is staffed with personnel who are trained to handle intelligence requirements for cyberspace, which is not always the case.

The main challenge in providing intelligence support for daily activities, the steady state, is communication between the MDT and supporting intelligence personnel. We heard from multiple intelligence organizations that intelligence requests from the MDTs are not always well crafted, either because the query is too broad ("What is the threat from adversary X?") or because the query focuses more on the system under threat ("Can an adversary exploit a vulnerability in a particular subsystem on my aircraft?"). This deficiency in specification likely reflects the fact that MDT personnel are unfamiliar with the intelligence process, reinforcing the need to work with OSS/IN to craft appropriate requests. On the receiving end, intelligence units and organizations will likely need to learn more about how to interpret and understand the ultimate objectives of the intelligence requests coming from MDTs. In our interviews, tactical intelligence units expressed their desire to transition from informal direct support to the MDTs to support the CDCC operating as a coordinating organization as this process matures.

Other intelligence support challenges arise when a specific cyber incident occurs. In our interviews, organizations responsible for coordinating incident response raised a notable concern: The organizational reporting chains for MDTs make them ultimately responsible to the wing and installation where they are assigned. This reporting chain is appropriate from a mission perspective, but it also means that the MDTs, despite a strong cyber flavor to their mission, sometimes do not report cyber incidents to the 616 OC in 16 Air Force.

Cyberspace Intelligence Tools and Products

Tools

Several intelligence organizations, including NASIC and the tactical units under the 70th Intelligence Surveillance Reconnaissance Wing, are adapting and revising their approach to intelligence products related to cyberspace, in part as a response to feedback from the field but also because of perceived inflexibility in some analytic processes. As one senior intelligence analyst noted, revising a comprehensive threat assessment can take months, and the MDTs and operational units often are looking for more-timely information. In non-cyberspace domains, the operational environment that ISR typically focuses on to collect data is often in Red and Gray space. In the cyberspace domain, there is an additional need to understand system susceptibilities to cyber operations in Blue space.

Our interviewees mentioned several tools that MDTs currently use. Most common were the CVA/H weapon system; Enterprise Logging Ingest Cyber Situational Awareness Refinery (ELICSAR), which is a big-data analytics platform from the company Leidos; and a commercial threat intelligence platform from the company Recorded Future. MDTs use a light version of the CVA/H weapon system. CVA/H is a defensive tool that performs threat assessment and system compliance on IT networks to identify and remediate network security vulnerabilities. It was designed to be deployed with CPTs. Interviewees discussed that the reason for providing the MDTs a light version of the tool was that the training and manning on an MDT were not enough to support utilizing the full tool. This tool is not a cyberspace intelligence tool, but it does provide the MDTs a capability to characterize Blue networks and provide inputs to make the networks less vulnerable. This is crucial to cyber threat intelligence because it allows threat analysts to focus their search.

ELICSAR services were acquired by the Department of the Air Force in 2019 from the company Perspecta. This tool aims to provide cyber situational awareness of networks with an enterprise-wide logging architecture. With this capability, the cyberspace intelligence analyst should have increased access to Blue network log data for processing, exploiting, and analyzing, in order to facilitate the search for cyber incidents in the network. For a typical all-source intelligence analyst, trained to process, exploit, and analyze Red systems, using such a tool focused on Blue systems will be an adjustment.

Lastly, some cyberspace intelligence units use a commercial threat intelligence service provided by Recorded Future.[40] This service provides access to cyber threat data from several

---

[40] Recorded Future, undated.

sources, such as technical data, open web sources, and dark web sources. Recorded Future can provide raw data as well as finished intelligence that is informed by its own research and other customer sources. Both the 35 IS and the 616 OC/ISRD stated that they use the service. The 616 OC/ISRD went on to state that the service is vital and that 616 OC/ISRD personnel believes that the cost of the service is less than what it would take to organically build a similar capability.

These three tools help intelligence analysts better understand the cyberspace operational environment. CVA/H allows for the understanding of Blue cyberspace terrain and enables MDTs to be better postured to defend it. ELICSAR allows analysts to hunt for incidents on a much larger scale than previously. And cyber threat intelligence feeds, such as is provided by Recorded Future, aid in the ability to be informed about threats from a diverse pool of sources.

Products

Turning from tools to products, we note that information-sharing processes between the MDTs and the intelligence organizations are still maturing, as described in previous sections. As a result, some formal products between the two are still being developed. Two main products were discussed during the interviews: RFIs and technical analyst reports.

The use of RFIs is standard in the intelligence community. Several of the intelligence organizations noted that the RFIs coming from the MDTs were difficult to respond to, for a variety of reasons. In many cases, the RFIs submitted by MDTs tend to be too vague and ask for information specific to Blue systems that traditional threat intelligence cannot provide. Other RFIs are too simple; some interviewees noted that the requesters should have the accesses needed to look up the answers themselves. There was no discussion of RFIs generating collection or production requirements; they appear to result in the MDTs being provided finished intelligence only if it was already available.

The second product mentioned as being useful to the MDTs was the technical analyst report. Produced by the 7 IS, technical analyst reports contain vulnerability assessments of specific Blue weapon systems. Although these products would likely be beneficial to MDTs, they are provided only to the customer that requests them. This is sometimes CROWS, but it is ultimately the program office that owns the information, controls classification, and determines to whom to release the report. One interviewee noted that they cannot even notify an MDT that such a report exists, much less share the information contained within it.

3. Recommendations for Mission Assurance at the Wing Level

> Mission assurance in and through cyberspace is not fundamentally an IT problem
> but a mission problem that requires a mission focus and approaches that go
> beyond what we have come to think of as traditional cybersecurity.
>
> — William D. Bryant[41]

In this chapter, we elaborate further on some of the findings in Chapter 2 and make recommendations to redress them. Recommendations are restricted to those that are relatively easy to implement yet likely to have significant positive impact. We return to the four strategic lines of effort of a wing-level organization:

- to defend its systems
- to respond to and recover from cyber incidents
- to maintain resiliency of its missions when systems fail
- to maintain sufficient situational awareness to make decisions to accomplish defense, response and recovery, and resiliency.

In the following sections, we step through tasks for each of these strategies, although we change the order of discussing each for ease of the flow of the arguments. Each section identifies current deficiencies in each task group along with potential remedies. The discussion builds on the observations of the last chapter, follows general guides from the academic field of organizational design, and acknowledges the mission assurance aspects of cyber mission assurance that place some equities on all personnel.

Not every task for each of these strategies can be done at the wing level. But the wing will play key roles in each. We will discuss which roles should be performed at the wing level and which should be performed elsewhere, supporting the wing. To justify arguments along these lines, we begin with a brief review of the literature of organizational design principles before turning to the four lines of effort.

Organizational Design Principles

Throughout the discussion that follows, a key question is which organizational units should be assigned key tasks in the above-mentioned strategy, within the Department of the Air Force and within each wing. We draw on two guiding concepts from the theory of organizational design to inform these recommendations. The first addresses suitable limits to the responsibilities of an organizational unit, its *span of control*. The second addresses the centralization or

---

[41] Bryant, 2016, p. 6.

decentralization of responsibilities according to the locus of relevant information, the *location of decision rights*. Both dimensions provide guides based on empirical experience of what has worked well in organizations.

Span of Control

As tasks become more complex, organizations tend to apportion them to specialized units. That is to say, experience indicates a limit to the range of complexity of tasks that a single organizational unit can effectively handle. So, if an organization is confronted with numerous complex tasks, it tends to divide these tasks among specialized units within the organization. If it has numerous such tasks, it often has numerous specialized units.[42]

Another factor that determines the size of units and the apportioning of tasks is whether an organization confronts significant competing demands. For example, if an organization needs to work some tasks quickly in response to the current environment and other tasks deliberately to plan for the future, these timelines conflict. The two circumstances also require different cultures. Therefore, the organization tends to apportion these tasks to different units. Assigning them to the same unit submits the unit to conflicting goals and operational tempi.[43]

We will use these principles to argue for limiting the specialization of tasks within individual organizations within wings.

Location of Decision Rights

Decision rights are often delegated to units (levels in the organization) where the necessary knowledge lies to make specific decisions. The alternative is for superiors at higher, more-centralized levels to elicit the necessary information to make the decisions themselves. When superiors elicit information to centralize decisions that could be made below them, they slow down decisionmaking and risk making poorer decisions if the transmission up the chain of command degrades the information. Removing decision rights from those at the locus of knowledge can also deflate morale because it signals lack of trust. In general, it has been argued that the more decisions can be delegated to the locus of relevant information, the better organizations function.[44]

That is not to say that all *authority* is delegated to those at the locus of the necessary knowledge. Decisions are normally guided by and constrained by direction from leaders higher in the chain of command. Commander's intent and rules of engagement are good examples of this kind of direction in the military. The intent of both is to provide enough guidance so that

---

[42] Bell, 1967; Blau, 1968; Ouchi and Dowling, 1974; Mintzberg, 1979, Chapter Eight.

[43] See Gaim et al., 2018, and references therein.

[44] Jensen and Meckling, 1992; Dessein, 2002; Marino and Matsusaka, 2005; Dobrajska, Billinger, and Karim, 2015. See also Coats and Rankin, 2017.

those with the immediate knowledge make the decision that central authorities would if they were in possession of that knowledge.

The information needed to make a decision could be multifaceted. The cyber realm presents circumstances in which key information for making some decisions is distributed among many parties. Technical knowledge of system details often lies lower in the organization, colocated with technical experts. But knowledge of operational and strategic implications for a decision, such as disconnecting an IT system or limiting operations on an aircraft, lies higher in the organization. These decisions require coordination mechanisms to find optimally informed choices. Although ultimate decision rights in such cases might be centralized, many actors contribute to the final decision.

Ubiquity of Equities for Cyber Mission Assurance

A theme we carry throughout this report is that cyber mission assurance is a mission problem, not a cybersecurity problem. Even if defending systems alone were sufficient, all personnel who interact with a system play a role in ensuring defense. They must know the proper procedures with respect to cybersecurity, practice sound hygiene, and be "sensors" for anomalous behavior.[45] But cyber defense of systems alone is insufficient. All four strategic lines of effort for the goal above are required. All four cannot be done without all members of wings playing a role.[46] *Cyber mission assurance will not, therefore, be achieved by assigning the responsibility for cyber mission assurance to a unit.* MDTs, as we argue below, have a key role, but they cannot perform all of the necessary roles. Each unit within a wing will have specific roles, but everyone will have some role. Absorbing this view is a necessary systemic cultural change in the Department of the Air Force, not just at the wing level.[47]

Organizational design principles also strongly indicate that cyber mission assurance cannot be assigned to a single unit. As we argue in the rest of this chapter, the number and complexity of the tasks, the enormous range of skills necessary to complete them, and the sometimes competing demands of tasks indicate that tasks should be apportioned among numerous units within a wing, not all assigned to one unit.[48]

Furthermore, all of these tasks need to be done during peacetime and during wartime, at home station and whenever deployed. To be effective at the time of the worst expected attacks, strategies must be able to meet wartime needs. During wartime, the scale and complexity of attacks can be expected to exceed the experiences of peacetime. Therefore, the capabilities of the wing must be scaled to wartime demands. Any needed reachback to other organizations must

---

[45] Snyder et al., 2017.

[46] Snyder et al., 2021, Chapter Five.

[47] Snyder et al., 2021.

[48] See Snyder et al., 2015; Valentine, 2018.

also scale with these demands and be available at the needed operational tempo. And the communications for the reachback must be resilient, even during adversary kinetic and cyber attacks.

In the following sections, we discuss tasks for each strategic line of effort and offer options for who might do them and how they might be done. We begin with situational awareness because it is foundational to the other three lines of effort.

Situational Awareness

Situational awareness helps prioritize the allocation of resources for cyber defense; informs detection of, response to, and recovery from cyber incidents; and guides cyber (mission) resiliency plans. We focus on two key aspects of situational awareness: mission structure and intelligence support. We discuss feedback and monitoring for management and command and control in the next chapter.

Understanding Mission Structure

Every system or process is, to some degree, susceptible to adversary cyber operations. A persistent challenge is that nearly every person or system can be an entry point for an adversary or an end target for cyber attack. Many leaders are left exasperated with this knowledge, unsure where to focus cyber defensive and cyber resiliency efforts. This quandary highlights the importance of understanding the structure of a mission. By *structure*, we mean all of the tasks necessary to carry out a mission; the interdependency of these tasks during mission execution, including temporal flow; and an understanding of which tasks are most critical; which systems, data, and personnel support each task, and how susceptible each is to cyber attack. The following are reasons for understanding the detailed mission structure:

- to be most effective in devising plans for the ability to survive and operate when under cyber attack and thereby achieve some level of cyber resiliency
- to prioritize cyber defensive efforts
- to define effective recovery plans
- to have better situational awareness of the state of cyber mission assurance, pre–, trans–, and post–cyber attack.

Each wing-level organization needs this level of understanding of its mission structure. The better this understanding is, and the more that it is kept current, the higher the quality of resiliency efforts will be and the better defensive measures can be prioritized.

One of the current expectations of MDTs is to describe the mission structure for their wing. For this purpose, they are trained in the FMA-C method. FMA-C is a method based on techniques that Nancy Leveson and colleagues developed for safety of complex systems at the

Massachusetts Institute of Technology.[49] Leveson's method emphasizes identifying unacceptable states that are to be avoided and uses concepts from control theory to avoid these conditions. In this frame, mission failures arise from poor application of constraints, lack of sufficient control, or inadequate feedback.

As a doctoral student with Leveson, Col William Young extended her methods to security, specifically to address cybersecurity from a mission perspective.[50] The current FMA-C used in the Department of the Air Force is based on this approach. We have reviewed the method and assess it to be sound and fit for purpose. Indeed, we have developed a similar method for the acquisition community that effectively takes the same approach.[51] Such techniques as FMA-C help an organization understand sources of potential unacceptable losses to its mission. When appropriately performed and used to guide cyber mission assurance efforts, such approaches as FMA-C can probably stymie well-devised attacks like Stuxnet.[52]

From our experience with our own method and observing the FMA-C method, we conclude that the constitution of the team carrying out the analysis is crucial for success. The quality of a mission map is only as good as the knowledge the team has of the nuances of that mission and the systems that support it. Knowledge of the method is necessary, but not sufficient, and the method is easier to learn than the nuances of a mission.

In our interviews, MDTs strongly expressed that, although they found FMA-C training useful for changing their mindset associated with the problem from one of cybersecurity to one of mission assurance, they found the training nearly useless for conducting a functional mission analysis of their wing's mission. To find remedies for this situation, we must understand the root cause of MDTs' struggles with mapping their wings' mission structures. We find the principal causes to be as follows:

- *Lack of realistic training.* According to our interviews with MDTs, the examples used in the FMA-C classroom are too simple compared with the real-world missions of wings. The examples were adequate to illustrate the power of the method and to convince the students of the utility of the approach. But students were not properly equipped to handle the significantly more complicated missions that they encountered upon returning to their wings.
- *Lack of exemplars.* MDTs have not been provided with exemplars of well-worked FMA-C analyses. Exemplars serve as an ideal, a template that the MDTs could use in mapping their wing's mission. At least three classes of exemplars would be initially needed: one for aircraft operations, one for an internet protocol network (such as the Distributed Common Ground System or AOC), and one for ICS.

---

[49] Leveson, 2004; Leveson et al., 2006; Leveson, 2011; Leveson, 2015.

[50] See Young and Leveson, 2014.

[51] Snyder et al., forthcoming; Mayer et al., forthcoming. Our approach also uses techniques from graph theory to help identify critical mission and system elements.

[52] Nourian and Madnick, 2018. The method used in this report was the System Theoretical Accident Model and Process, which is fundamentally the same method as FMA-C. See Young and Leveson, 2014.

- *Lack of support.* The training gives MDTs several days of classroom theory and practice. Then, they are on their own. When they confront the difficulties of mapping their unique missions, no reachback support or field teams are available to assist them.
- *Lack of skilled practitioners.* Those tasked with doing this analysis of the wing's mission have insufficient backgrounds and skills to understand the wing's mission or how it fits into the joint war fight. The members of the team are junior, generally led by a lieutenant. Most teams have one noncommissioned officer, and the remaining members are frequently below that grade. Most have training and experience in installing and maintaining computer networks. They do not generally have security backgrounds or any direct experience in operations.
- *Lack of quality control.* Once MDTs conduct an FMA-C analysis of their wing's mission, there is no evident formal check of the quality of their work. They do not know how well they have done, and commanders are not, in general, sufficiently trained in the method to assess the quality.

The ultimate recipient of FMA-C (or similar) analysis is the wing commander. The point of the analysis is to inform the commander's decisions for cyber mission assurance. Hence, a further deficiency is that wing commanders have not been trained in what to do with situational awareness of this type. How, exactly, should they put these results to use in prioritizing cyber defense, devising cyber resiliency strategies, planning for recovery, and assessing the cyber mission assurance of their wing's mission? No specific training is provided to wing commanders in preparation for the position or after taking it. The interviews that we conducted revealed that the quality of the use of FMA-C analysis in a wing depends on the previous experiences and knowledge of the commander, which vary considerably.

Although there are nuances for practical implementation, the remedies for these problems are straightforward to state:

- enhanced FMA-C training that is more realistic
- advisory support for wings after training
- more-qualified teams within the wings performing mission mapping
- some training for wing commanders on how to effectively use FMA-C analysis.

Enhanced Training

An ideal training program for a wing FMA-C team would entail working an example mission pertinent to the wing's mission that possesses comparable complexity. But such an exercise would take months, exceeding the available week for this training. At best, within the week allocated, instructors could help the wings understand the types of complexities they are likely to encounter when returning to their units. This additional instruction should be tailored to the nature of the wings' mission (e.g., aircraft-focused, command and control–focused). The training should provide them with additional guidance for how to handle these complexities.

Advisory Support

Students will not be able to tackle in class a problem of the complexity of their wing's mission. And to some degree, each wing has a unique mission structure. Hence, a weeklong training session is unlikely to adequately prepare a team to perform FMA-C analysis on their wing's mission. But they can be introduced to the practical difficulties and pitfalls that commonly befall an effort at a wing.

This circumstance suggests that the course be seen as foundational knowledge for the wing, not definitive training for FMA-C. Further advisory support would be extended to the wing teams when they return to their units. A central organization, associated with those teaching FMA-C, could deploy as field teams to support wings during mission mapping. It would also be beneficial if some organization served as a quality-control check on the work. That organization could further archive mission maps to serve as guides and templates for future efforts by similar wings.

Appropriately Qualified Teams

A holistic assessment from an operational perspective is central to mapping the mission. Essential expertise to map the mission lies within the Operations Group of the wing. But all groups within the wing need to participate. For successful continuity of wing operations, the maintenance and support activities must also function. Mission mapping includes every part of the wing's mission, and, therefore, expertise from every part must flow into the mapping.

Although every element plays a role, one part of the wing needs to take the lead in coordinating and compiling the mission map by FMA-C. Currently, this role is assigned to the MDT. But FMA-C is about mission assurance, not what is commonly called *cybersecurity*. The ultimate aim of FMA-C analysis is to identify the unacceptable hazards to the mission. The right skills and background are those of understanding how all of the systems and processes of the wing (and, to some degree, outside the wing) combine to perform the mission. A background in communications is not suitable. The commanders and Operations Group must be involved in making these judgments, not just the personnel in the MDTs.

The role of the team doing the mapping, therefore, is one of a facilitator; the team builds this map of the wing's mission by consulting with the rest of the wing. To perform this facilitation role effectively, the team needs to have more operationally knowledgeable personnel and sufficient status in the wing to gain the cooperation of other units in the wing. To understand the nuances of what is truly mission critical and how the wing might operate when under the duress of a cyber attack, the personnel should have more experience and higher rank than in current MDTs.[53] As we also argue in the section on cyber defense, the span of control of the MDTs gets very wide when they are asked to execute such roles as FMA-C and cyber defense.

---

[53] See Mayer et al., forthcoming, for an extended discussion of how a mapping effort can be effectively executed.

For these reasons, we recommend that the FMA-C role be removed from MDTs and that the role of mission mapping via FMA-C be assigned by the wing commander to a unit with more-experienced leadership and with more operational knowledge.[54] But this unit must still work with the rest of the wing (and perhaps the host wing) to perform holistic FMA-C.

Training for Wing Commanders

Wing commanders need some training regarding FMA-C. They need to know what FMA-C is meant to accomplish so that they know good analysis when they see it, and they need to be taught what FMA-C can do for them. It can help them determine priorities for which systems to emphasize for cyber defense. They should be taught how FMA-C can help them better plan for cyber resiliency and how it can give them better situational awareness of their wing's cyber mission assurance. An appropriately shaped module could probably be given to a wing commander in a customized, one-to-two-hour session.

Summary of Recommendations

Assign responsibilities for FMA-C to units within wings with at least O-5-level leadership. The team should contain considerable operational knowledge and should reach out to every element of the wing.

Remove direct responsibility for executing FMA-C from the MDTs. MDTs should participate in FMA-C as expert sources, but they have neither the proper level of experience nor the operational perspective for this task and already have a wide span of control.

Create a temporary capability to support wings while they perform FMA-C via reachback to a center of excellence for FMA-C or field teams that can assist wings. Doing an FMA-C analysis for the first time requires a lot of work and expertise. When wings are doing this work for the first time, they need assistance from experts in the method to help them as they work through the challenges of their unique mission. When all wings have done an FMA-C analysis, this support capability could be disestablished, since maintaining a mission map is not nearly as challenging as creating it for the first time.

Check the quality of wing FMA-C analysis for its adherence to the methodology by the FMA-C instructional team, and check the quality of the FMA-C analysis for accurate representation of the mission by the wing commander.

Create training for wing commanders on how they can best use FMA-C analysis.

Intelligence Support

The Department of the Air Force has begun to put in place the organizational structures, develop the analytic platforms, and refine the necessary processes and products for providing cyberspace intelligence support to operational units. These efforts are works in progress. The

---

[54] Some MDTs made this same recommendation.

recommendations provided here should assist in that process and ensure that appropriate intelligence support can be provided to the MDTs as the MDT initiative matures.

Ensure that the process for requesting and coordinating intelligence support to wings is defined, published, and widely distributed. As noted in the last chapter, tactical cyberspace intelligence units have often received direct requests for intelligence support from MDTs, but this approach can lead to redundancy if multiple organizations request similar intelligence and can quickly overwhelm a unit. The current informal process based on relationships may work for the pathfinder MDTs but is unsustainable as the enterprise grows. The CDCC construct, which is emerging, can not only address the coordination of these requests but also answer questions or similar inquiries that the CDCC has seen before. However, for the construct to be effective, MDTs need to be fully aware of how to participate in the process for both steady-state daily activities and cyber incident response.

Establish best practices and guidelines for cyber RFIs. A common complaint across all of the organizations we interviewed is that there are insufficient information and products available on realistic cyber threats to Blue systems. MDT personnel need tactical intelligence that is actionable, but they struggle to write appropriate intelligence requests, asking questions that are either too broad or too specific to Blue systems. Providing guidelines to MDT personnel for how to write appropriate RFIs and training at least some of them in the basics of intelligence tradecraft would go a long way toward addressing this issue until the intelligence billets are filled in MDTs.

Develop a complete curriculum for intelligence analysts who support wings for cyber mission assurance. We heard across units that all-source intelligence analysts, the most common AFSC in OSS/IN units, receive modest education in cyber fundamentals—often just a few weeks— before or shortly after joining a unit providing cyberspace intelligence analysis. Rarely did we hear of any continuing education. One senior intelligence analyst noted that the combination of skills needed—intelligence tradecraft, engineering expertise, and sometimes knowledge of acquisition programs—is difficult to find in any one person. Another senior analyst noted that it was critical that all-source analysts were skilled in the intelligence tradecraft first and then learned the cyber fundamentals. As analysts gain experience in their roles, they will also become aware of knowledge gaps that continuing education and training can address.

Conduct a full assessment of the long-term cyberspace intelligence needs across the force. The deployment of MDTs across the force will create increasing demands for cyberspace intelligence. Although some intelligence units appear to be meeting current demand despite unfilled billets, that state of affairs may not be sustainable over the long term. At the same time, as platforms like the ELICSAR are populated with more data and analysts learn to use them to "pull" the analysis they need, there may be efficiencies that can be realized. The Department of the Air Force should assess how well the current and planned intelligence enterprise is postured to meet expected future demand for cyberspace intelligence to support mission resiliency, both during peacetime and during a potentially higher operational tempo during wartime.

Defense

Cyber operations directly affect systems. So, it is systems, not missions, that are the objects of cyber defense. Cyber defense includes such activities as discovering and fixing vulnerabilities, monitoring system behavior for anomalies, and detecting suspicious activity.

A wing uses a number of diverse systems to carry out its missions. As described in Chapter 1, for the purposes of cyber defense, it is useful to categorize systems into three bins of comparable characteristics: cyber-physical systems (such as aircraft), ICS (such as controllers for distribution of power and fuel), and IT systems (such as the Non-Secure Internet Protocol Router Network [NIPRNet], SIPRNet, and Falconer). The architectures, protocols, operating systems, and other configurational details differ widely among these three groups. Techniques and constraints on monitoring these systems with sensors can also differ considerably. The missions of most wings depend on systems in all three categories, but the skills needed to defend them differ.

Adversaries can choose to attack any one or all of these types of systems. Cyber defense at a wing must adequately cover all three cyber system categories. Attempting to defend everything at the same level of effort would be a misuse of resources, as not everything is of equal importance and not everything is equally susceptible to attack. We take up how to establish that importance in the section on situational awareness.

Cyber-Physical Systems

Issues

Cyber-physical systems—such as aircraft, missiles, and satellites—require different skills and tools to defend than IT systems. Cyber-physical systems differ from IT systems in significant ways. In these systems, hardware and software integrate to form a synergistic entity. Unlike most IT systems, the timing of the execution of software must synchronize with hardware states. Latency in software execution can cause system failure and safety hazards. Any sensors for cyber defensive monitoring must not interfere with this coordinated orchestration. Sensors must also, in the case of an aircraft, meet constraints of weight, balance, power consumption, and heat dissipation.

Yet most cyber-physical weapon systems also exchange data at their margins with IT systems using internet protocols. For aircraft, for example, it is common for the cyber-physical systems to connect with maintenance or test equipment. These systems often run commercial operating systems and exchange data, directly or indirectly, with internet-connected systems. Defending cyber-physical weapon systems, therefore, requires knowledge and experience of the cyber-physical weapon system as well as IT systems.

Although technical knowledge of IT systems applies across many different systems because of common protocols and software, the same does not apply to cyber-physical systems. It is true that some technical aspects are similar or identical across cyber-physical weapon systems because of standardization. The military standard 1553 data bus on aircraft, commercial aircraft data buses, and common test equipment are good examples. But individual components, such as line-replaceable units, sometimes run bespoke software on specialized operating systems or without an operating system at all. These components sometimes differ even within a single aircraft. They generally differ among aircraft. Technical knowledge of one cyber-physical weapon system does not prepare personnel sufficiently for defending another cyber-physical weapon system.

Gaining a comprehensive understanding of a cyber-physical weapon system is sometimes not possible. For some systems, the software might be proprietary to a vendor. Proprietary data limit the ability of Department of the Air Force personnel to access some parts of the system, maintain sufficient situational awareness of the entire system for cyber defense, or both. The consequences of this limitation are compounded by the fact that it is the most-recent systems, such as the F-35A, that contain significant proprietary software, and it is these same systems that are the most dependent on cyber functionality—and, therefore, the most susceptible to cyber operations and the most in need of careful defense.

Wings do not, in general, possess tailored tools for cyber defense of their cyber-physical weapon systems. The generic tool supplied to MDTs and for which they are trained is CVA/H. CVA/H is designed for IT systems. As such, it can be suitable for the IT-configured margins of a cyber-physical weapon system, but it is not designed for the peculiarities of the cyber-physical weapon system core. Indeed, a number of wings with non-IT systems explained such challenges during our interviews.

Although the locus of knowledge of the technical engineering details of a cyber-physical weapon system lies at the corresponding PMO, the day-to-day knowledge of the normal patterns of behavior of the system and the consequences of system failure to the mission lies in the operating wing. The operating wing also has continuous access to the system, at home station and deployed, which the PMO does not.

Recommendations

With those observations, we now discuss options for improving wing-level cyber mission assurance via cyber defense of cyber-physical weapon systems.

Assign cyber defense of cyber-physical weapon systems to specialized teams dedicated to individual system types. The complexity and range of the tasks for defending cyber-physical weapon systems indicate that assigning cyber resiliency duties or other non-system-related duties to these units risks poor performance on some or all of their assigned tasks. MDTs were largely formed with the task of defending weapon systems (despite the name *Mission* Defense Team). MDTs would be a logical choice for the role of defending cyber-physical weapon systems. But MDTs in policy and practice have many other responsibilities that distract from defending cyber-physical weapon systems. One of those responsibilities is mapping the mission using FMA-C to identify mission "hazards" and to help identify priorities for defense and cyber resiliency of the

42

mission. This task, which we discuss in detail in the section on situational awareness, requires a comprehensive understanding of the wing's mission, the full array of systems that support it, and the architecture of how those systems support the mission. That is a very different skill set than the one required for cyber defense of cyber-physical weapon systems. We recommend limiting the role of MDTs to system defense.

Do not rely on CPTs to augment MDTs' defense of cyber-physical weapon systems. Although CPTs currently bring a higher skill level to cyber defense and incident response, they do not, nor can they practically, learn and maintain the necessary expertise in each cyber-physical system. MDTs should, in the long term, develop into teams with all of the necessary skills to defend their systems. They will need assistance in various forms to respond to and recover from a cyber incident, which we take up in the section on response and recovery.

Develop training for MDTs tailored to their cyber-physical weapon systems. MDTs that defend cyber-physical weapon systems need a high level of expertise, given the complexity of these systems. They need training for these weapon systems in a similar way that aircraft maintainers get training specific to their weapon systems. This training should augment training in defending IT systems, which is needed to defend the margins of the systems.[55]

Ensure that MDTs dedicated to defending each cyber-physical weapon system qualify for and possess the appropriate security clearances and any special accesses needed. We understand that the point of special accesses is to limit the number of individuals with access to very sensitive information to reduce the risk of insider threats. In particular, as a result of lacking the appropriate accesses, MDTs may not even be aware of vulnerability assessments that have already been conducted on their weapon systems. Without appropriate accesses, MDTs cannot adequately defend their assigned systems.

Assign strategies for defending cyber-physical weapon systems and for developing appropriate tools to the corresponding PMO. A center of excellence is needed for each cyber-physical weapon system to identify and correct cyber susceptibilities, to understand the effects on a system from various cyber operations, and to consult with prime contractors for additional engineering advice when needed. The locus of this knowledge and the authority to consult prime contractors lie at the PMO. Rather than re-create these capabilities elsewhere, we recommend that the PMO, with active participation of the applicable AO(s),

- develop technical orders for the monitoring and defense of its cyber-physical weapon systems
- be given responsibility to supply tools, including any sensors, for the cyber defense of the systems, in lieu of generic, one-size-fits-all tools, such as CVA/H.

---

[55] The specialized training and sustaining the requisite skills at the wing level for these MDTs will be a challenge for cyber-physical weapon systems. How to manage this problem will vary according to the type of weapon system and the fleet size. Solutions to these problems are beyond the scope of this report.

MDTs for cyber-physical weapon systems would then carry out the technical orders using tailored tools provided by the PMO and approved by the applicable AO(s). This process would mimic aircraft maintenance, in which the PMO issues technical orders and the wing executes the work. An MDT will encounter issues from time to time (other than cyber incidents, to be discussed later) that technical orders do not adequately address. The MDT will need a process—modeled on maintenance "107 requests"[56]—that requests assistance from the PMO. This overall process of technical orders and "cyber 107 requests" would standardize and codify cyber defense of cyber-physical weapon systems.

There are limits to the reachback capabilities that a PMO can provide. PMOs are not staffed to operate on a 24-hour, 7-day basis. Nor are they staffed to surge to meet the additional capabilities during wartime. The role of the PMO should, therefore, be restricted to technical orders and cyber 107 requests. We will discuss other operational-type reachback capabilities for wings in the section on response and recovery.

Industrial Control Systems

Issues

ICS is a collection of sensors and programmable logic controllers connected to form a system to manage infrastructure facilities, such as relays for power and valves and pumps for fuel. ICS collects data for system monitoring, facilitates maintenance, and controls the state of the infrastructure. Increasingly, these systems are being connected via internet protocol networks, including wireless connections. These networks are local-area or wide-area networks increasingly connected to business networks for convenience of operations, separated from them only by a firewall or a "demilitarized zone."[57]

ICS is challenging to defend. Like cyber-physical systems, ICS often run proprietary operating systems and software. The proprietary components limit the access and situational awareness that the Department of the Air Force can maintain. Furthermore, market forces have been a poor driver for cybersecurity of ICS, resulting in vulnerable systems by design.[58] The physical location of ICS devices, widely distributed in locations in the field without easy access, hinders cyber defense. In addition, many ICS devices must operate continuously and therefore cannot be shut down on demand for software patches.[59] They are susceptible to cyber attack and are difficult to defend. Yet ICS underpins most capabilities at a wing. The Stuxnet operation was an attack on an ICS.

---

[56] Secretary of the Air Force, 2011.

[57] In network security, a *demilitarized zone* is a layer separating a trusted inner network from an untrusted external network.

[58] Knowles et al., 2015; Nazir, Patel, and Patel, 2017; Ashibani and Mahmoud, 2017; Upadhyay and Sampalli, 2020.

[59] See Gonda, 2014.

Organizational issues present additional challenges for defense. At home station, ICS falls under the responsibility of the host unit (which might be an air base wing). That host unit must maintain excellent lines of communication with all of the tenant units to understand which mission elements depend on which ICS devices to ensure prioritization of defensive efforts. But many ICS extend beyond the installation boundary and are owned and operated by commercial firms, complicating this coordination.

Deployment presents an additional challenge. Because tenant units do not have ICS responsibilities at home station, they do not have organic capabilities for ICS cyber defense when deployed. As at home station, some of these ICS capabilities will lie outside the base and be under host nation or commercial control.

Recommendations

Establish a center of excellence for ICS cyber defense. This organization would, like a PMO, issue technical orders for cyber defense of ICS. It would keep current on best practices of defending ICS, including such new techniques as behavioral anomaly detection, including via machine learning.[60] From this base of knowledge, it would develop and supply appropriate tools and sensors to wings. It would also create the training program for wing-level teams defending ICS.

Create MDTs dedicated to cyber defense of ICS. Like the MDTs dedicated to cyber defense of cyber-physical weapon systems, these ICS MDTs would have no other responsibilities. They would be trained and skilled in ICS cyber defense. These MDTs would follow technical orders issued by the ICS center of excellence.

Develop UTCs for ICS cyber defense. The U.S. Air Force will need to develop and maintain UTCs for ICS cyber defense akin to combat communications units. The capacity of such teams in the U.S. Air Force would be sufficient to defend ICS at home station and supply UTCs for deployment to cover around-the-clock operations.

Information Technology Systems

Issues

Communications networks—such as the NIPRNet, the SIPRNet, and other networks—play critical roles in a wing's operations. The ultimate responsibility for defending these IT systems is assigned to 16th Air Force under authorities from U.S. Cyber Command.

Prior to the EITaaS initiative, the Communications Squadron within a wing played a significant role in defending IT networks inside an installation. The EITaaS initiative now effectively outsources the cyber defense of these IT networks. We learned during our wing-level interviews that this outsourcing presents two challenges to wings when at home station. First, the

---

[60] Bhamare et al., 2020; McCarthy et al., 2020.

success of cyber defense of IT hinges on the quality of the contract language. The government will get only what it has contracted for, and deviation of the measures of performance for the contractor from mission objectives of the wing will result in mission fragility when under attack. Given the uncertainties in the cyber realm, writing this exact language is difficult. Second, under EITaaS, the wing loses a certain degree of situational awareness of the configurations and security of the networks that support its mission. That loss impedes the ability to map the mission, discussed later, and develop robust plans for cyber resiliency.

For outsourcing to be successful, the U.S. Air Force will need to compensate for the loss of airmen experienced in setting up IT networks, which are needed for deployment, and in higher demand with distributed operations (maneuver). The U.S. Air Force will need deployable teams to defend these networks when deployed as well. Those combat communications teams will need to have the skills and capacity to meet these demands.

Recommendations

The focus of this report is on wing-level aspects of cyber mission assurance other than defense of IT systems. Although the IT systems form a backbone for wing communications, they are out of the scope of this report. So, we offer no specific recommendations to mitigate these issues other than to note that the challenges that we have listed will need to be addressed to enable a holistic approach to cyber mission assurance.

Response and Recovery

Issues

Responding to and recovering from a cyber attack is an integral part of cyber defense of systems and cyber resiliency of the mission. We will not dwell on aspects of response and recovery that are integral to defense and resiliency. We focus here on the processes for reporting and making useful meaning out of cyber incidents to inform decisions on how to respond and recover.

A *cyber incident* is defined by the Committee on National Security Systems as follows:

> Actions taken through the use of an information system or network that result in an actual or potentially adverse effect on an information system, network, and/or the information residing therein.[61]

It is defined similarly by the White House as follows:

> An event occurring on or conducted through a computer network that actually or imminently jeopardizes the integrity, confidentiality, or availability of computers, information or communications systems or networks, physical or virtual infrastructure controlled by computers or information systems, or information

---

[61] Committee on National Security Systems, 2015.

46

resident thereon. For purposes of this directive, a cyber incident may include a vulnerability in an information system, system security procedures, internal controls, or implementation that could be exploited by a threat source.[62]

These definitions of a cyber incident do not reflect the uncertainties that surround cyber anomalies. Three states can be defined:

1. Something looks suspicious, but not enough is known about its potential harm. One example of this state might be that monitoring of a system indicates that it has some attribute that is out-of-bounds of previous experience, perhaps a file that is new and unaccounted for or traffic in a network that is unusual. Another example might be an airman noticing that data look unusual and might have been corrupted. In these cases, the observation might indicate an attack, or the anomaly might be from other causes, or it might not be an issue at all.
2. Something is definitely problematic, but not enough is known to establish that it is harmful or to what degree. An example might be malware that is discovered on a system, but it is not known whether the malware can harmfully affect that system.
3. Something is definitely harmful. In this case, some cyber operation is confirmed and some part of a system or mission is at risk.

The term *cyber incident* does not clearly distinguish these nuances.

The three states might not be the exact right ones for triggering various actions and notifications. It is beyond the scope of this report to define exactly what the categories of cyber incidents should be for reporting. But some more-nuanced categories should be defined for when a cyber anomaly should be reported to a wing commander, when it should be reported outside the wing, and who should be notified. These should be defined precisely enough to be reasonably unambiguous and should be consistent across units. Our interviews indicated that these nuances do not exist in reporting, and reporting is therefore ad hoc both within and among wings.

Initially, it can be ambiguous whether or not a cyber incident, in the broader sense, is one of concern. The goal of incident response that we want to address here is identifying what, if anything, is really at risk from the cyber incident. The goal of identifying this information is to guide decisions on how to stop any cyber attacks that are in progress and prevent further propagation of the attacker's effects.

The cyber incidents of most concern are those that are potentially damaging to U.S. operations during wartime. The Department of the Air Force will need to prepare as though these incidents will be a persistent campaign, not a single attack, and as though the adversary will attack multiple targets. Sufficient capacity will be needed to address these incidents across the wide range of system types. Response time will be important, not only to mitigate the effects on missions but also to decide and act in time to stop any attacks and combat ongoing adversary cyber operations. Quick response helps prevent Red from getting inside Blue's OODA loop.

To this end these activities are critical:

---

[62] The White House, 2016.

- prioritizing the assessments of cyber incidents and establishing how quickly each needs to be performed
- assigning each incident to an assessor or assessors and distributing incident information to any other organizations for situational awareness
- assessing incidents
- disseminating findings from the assessments
- compiling, archiving, and disseminating lessons learned.[63]

During a large-scale attack during wartime, the volume of cyber incidents might be too high for the resources available. Should that situation arise, there will be a need to prioritize and triage the efforts. Which incidents have the greatest risk for significant mission impact? How vital are those missions? Which incidents are the most time-critical to address? How should limited resources for assessments be allocated?

After any needed prioritization, each incident needs to be assigned to suitable experts for forensics and assessments. Cyber forensics and assessments of a cyber incident can be highly technical. They might require the participation of more than one organization and multiple tools. Organizations for assessment might include PMOs (who might consult vendors), the intelligence and counterintelligence communities, CPTs, and other centers of excellence. The resource capacity for these activities will be finite, reinforcing the need for prioritization and triage in the form of command and control of the allocation of these resources. Furthermore, during wartime, these resources will need to be operationally responsive. They will need to be able to operate at all hours and at the operational tempo of the warfighter. Findings from these assessments will then need to be distributed to organizations with a need to know.

Finally, some archive should be kept of cyber incidents by system type to maintain a record of previous issues, much like a medical record, and to facilitate future assessments.

Existing policy for incident response is not yet mature enough to prescribe all of these activities across cyber-physical systems, ICS, and IT systems and assign them to appropriate organizations with well-understood and well-exercised processes.[64]

In some simple cases, cyber incidents can be identified and assessed within a wing as benign. But in many cases, information about the incident should be shared outside the wing to ensure that the wing has not missed something and that patterns of attacks across units can be identified. So, most of the activities just described will happen above the wing level.

Interviews with the wings revealed a lack of consistency in terms of what constituted a reportable cyber incident, what to report, to whom, and which reporting systems to use. Most wings expected to handle each case individually, on an ad hoc basis. This state is understandable. The Department of the Air Force has yet to develop fully functional roles and responsibilities

---

[63] Snyder et al., 2017, Chapter Four.

[64] Cichonski et al., 2012; Bartock et al., 2016; Committee on National Security Systems, 2016; Chairman of the Joint Chiefs of Staff Manual 3150.07E, 2013; Chairman of the Joint Chiefs of Staff Manual 6510.01B, July 10, 2012, Directive Current as of December 18, 2014; Air Force Instruction 17-203, 2017.

and command and control for cyber incident response. One reason for that deficiency is lack of experience in serious cyber attack, such as might happen during war. The experience base has not yet accumulated to the point where it is understood what works well and, therefore, how to direct policy.

Recommendations

The reporting thresholds and processes cannot be fixed at the wing level before they are established throughout the Department of the Air Force. At the core of the central activities described in the previous section—prioritizing, assigning, assessing cyber incidents, and disseminating and compiling findings—is command and control. But this command and control needs to integrate across the entire operational Department of the Air Force and system types (cyber-physical, ICS, and IT). It is a bit beyond the scope of this report to propose a complete assignment of roles and responsibilities for cyber incident response. A recommendation is to write more-specific policy for cyber incident response that assigns these roles and responsibilities along with clearer processes.

However, the principle of unity of command suggests a single command center within the Department of the Air Force to coordinate these activities. This center would need to be operational 24 hours per day, able to surge to wartime needs, and able to handle compartmented and special-access information. It would need to operate like an AOC.

Resiliency

Issues

The ultimate goal is to assure a wing's mission. Defense of systems and their recovery after cyber attacks is one set of means to this end. Some functional restoration will require system recovery. For example, some maintenance requires certain diagnostic and test equipment. But some functions can be continued, albeit sometimes with some degradation, by some other means. Those means might be using other systems as substitutes, altering procedures, or using some other adaptation. An example would be using alternate communications systems when primary systems are down. We call these collective means *resiliency*.

Resiliency complements system recovery. Resilient strategies provide ways to continue the wing's core mission elements during and after an attack other than by defending and recovering systems. This route creates and exploits resiliency of the mission to adapt to the attack and adjust operations to work around loss of data, corruption of data, loss of communications systems, or damage or destruction of systems via cyber, or any, attack.

Resiliency is about establishing a mission architecture that can absorb attacks and recover from them to assure the wing's mission. It is also about exploiting that architecture—by knowing how to adjust operations to meet the challenges of cyber attacks. System recovery can be part of that, of course, but, here, we focus on the structure of the mission—how all of the resources and processes for mission accomplishment interact and how this big-picture view of the mission enables resiliency. In doing so, it also identifies systems, such as the diagnostic and maintenance systems mentioned earlier, for which failure has no clear avenues of redress other than system recovery. These are areas of critical fragility that should be known as part of situational awareness.

Every part of a wing plays a role in resiliency because every part of a wing contributes to the wing's mission. Resiliency cannot be done by a single unit. Achieving resiliency might involve using a different system as a substitute for a disabled one, such as a backup communications system. It might be an alternative means of carrying out a supply process when the standard data systems have been corrupted. It could be a backup for electronic maintenance technical orders. It could be an alternative means for mission planning when supporting computers are unreliable.

Resiliency is a problem-solving activity. It involves personnel at each part of the wing affected by an attack identifying alternative means for carrying out the mission and implementing those means. Some of this problem-solving will be done in advance of an attack. But because all attack effects cannot be anticipated, responses cannot be fully scripted, so some problem-solving will be done after an attack. This attribute of finding adaptive solutions to carry out a mission under duress that look different from the way the mission normally operates is commonly observed during disaster recovery of infrastructure.[65]

The underlying knowledge needed to guide problem-solving for resiliency is the mission map, which is the intended product of FMA-C analysis. The mission map forces personnel to understand the big picture from a mission perspective. It places systems and processes in an operational context. The mission map reveals elements that share a common cyber susceptibility and whose failure would jeopardize the mission—that is, critical mission elements.[66] It helps identify where to prioritize efforts for defense and resiliency.

Everyone in the wing has a role in cyber resiliency, but some unit needs to coordinate these efforts to ensure that critical elements are identified and that resiliency plans are devised and exercised for them. Experience from disaster recovery indicates that the role of this team is to set goals and coordinate and facilitate the efforts of the rest of the organization. The team needs to get out of the way of the wing members in each functional area who have the knowledge of what to do.[67]

The skills that this cyber, or mission, resiliency team needs are distinctly different from those of the MDTs defending systems. At the system level, activities revolve around a nucleus of aspects of the problem that is composed of technical facets of the system, the system configuration, how humans interact with the system, the system vulnerabilities, and similar

---

[65] Roe and Schulman, 2016.

[66] Snyder et al., forthcoming.

[67] Roe and Schulman, 2016.

details. MDTs need detailed technical understanding of their assigned systems and familiarity with what constitutes normal behavior of the system.

At the mission level, resiliency must take a holistic perspective of how all of the systems, personnel, and processes work in concert to achieve a mission. The mission level is not about details, but about connections, integration, and the overall structure of the mission. It must keep the mission objectives central. Those working at the system level for cyber mission assurance look up from the system level to the mission level. Those working at the mission level look down to the myriads of systems and other components of the mission. Such different perspectives and skills are generally assigned to different organizational units.

Recommendations

A wing needs to continue its mission despite cyber attacks. Some of the efforts for cyber mission assurance lie at the level of system defense and system restoration, and some lie at the level of mission resiliency and ensuring and restoring the mission other than by system defense and system restoration. As described in the previous section, at the system level, activities revolve around a nucleus of aspects of the problem that is composed of technical facets of the system, the system configuration, how humans interact with the system, the system vulnerabilities, and similar details.

At the mission-resiliency level, activities take a holistic perspective of how all of the systems, personnel, and processes work in concert to achieve a mission. The mission-resiliency level is not about details but about connections, integration, and the overall structure of the mission. Those working at the system level for cyber mission assurance look up from the system level to the mission level. Those working at the mission-resiliency level look down to the myriad systems and other components of the mission.[68]

Working at the system level and working at the mission-resiliency level require different skills and foster different cultures. From the principles of organizational design, it makes sense to assign tasks at these two levels to different units. We therefore recommend separating the roles of cyber defense of systems (assigned to MDTs) from those of cyber resiliency (assigned to a different unit trained in FMA-C methods).

Figure 3.1 shows the three main strategic groups of activities for cyber mission assurance in relation to the system-mission perspective.

---

[68] Roe and Schulman, 2016, pp. 43–50; Klein, 1998, Chapter Eight.

Figure 3.1. Cyber Mission Assurance Strategies Across Systems and Missions

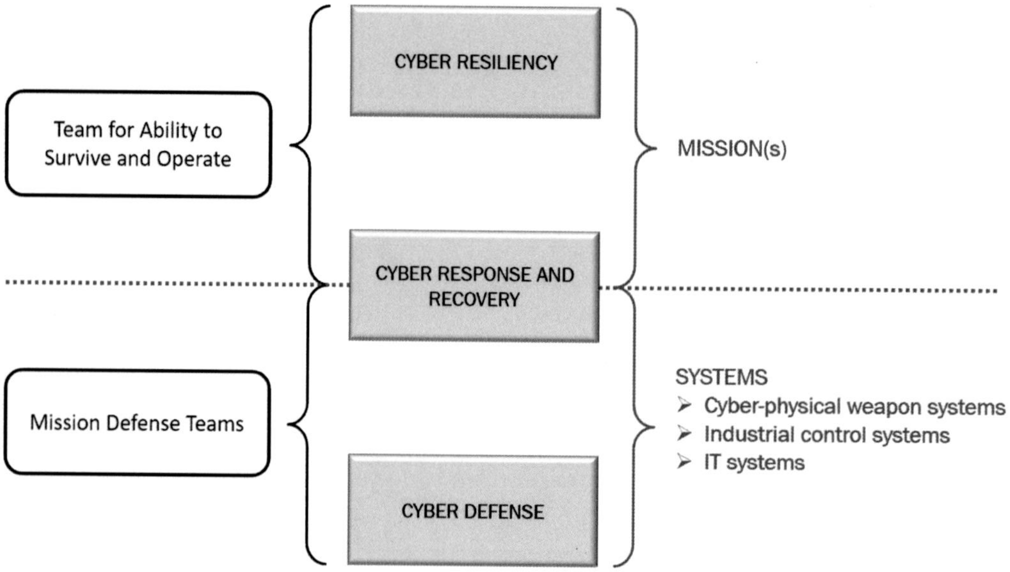

As shown at the top of Figure 3.1, mission resiliency to cyber operations resides at the mission level. Resiliency is about finding and implementing work-arounds, substituting one resource for another that has failed, or replacing a connection that is lost with a different one. Cyber resiliency is not about fixing failed systems or processes. It is about ensuring the overall mission by adjusting which elements come together for the mission and how they are connected. It accepts system failures and works around them by adapting. As we argued earlier, and reinforce here, we recommend assigning the role of managing resiliency to a single team with this cultural focus and skill set.

At the bottom of the figure, the system level comprises the full set of systems the wing needs for its mission(s). These are the cyber-physical weapon systems, the ICS, and the full spectrum of IT systems. This level is about defense, including preventing system failure. The practitioners at this level do not need to understand every other system and how all of the systems fit together for their mission. The job of defense necessitates detailed knowledge and continuous attention. As we argued earlier, we recommend assigning these duties to specialized teams for ICS and cyber-physical systems, perhaps a team per cyber-physical system type, depending on the complexity of the systems and the similarities across them.

The box depicting cyber response and recovery sits in the middle of the figure, straddling the horizon between the system and mission levels. When cyber attacks happen, the mission level needs to find solutions for how the overall mission architecture can adapt. The system level needs to try to restore system capabilities, confine malware and cyber effects, and prevent further attacks. So, both teams play significant roles in response and recovery.

To summarize, we recommend that

- teams for the ability to survive and operate in a cyber-contested environment

- perform and maintain mission maps for the wing using FMA-C or a similar method
- recommend preemptive adjustments to mission architecture for enhanced resiliency
- devise and exercise plans for mission continuity during and after cyber attacks
- advise the commander, during and after cyber attacks, of potential courses of action for mission continuity.

- MDTs[69]
  - defend systems, such that some team defends each system type
  - respond to cyber attacks of systems and recover system capabilities thereafter.

The teams for the ability to survive and operate in a cyber-contested environment could be placed at the discretion of the wing commander, as a direct reporting unit to the commander, as part of the command post, as part of the wing operations center, or as part of the mission planning cell, for example. These teams would not necessarily need to exist as continuously operating entities. Mission maps take time to create, but much less time to maintain. Membership of teams for the ability to survive and operate in a cyber-contested environment could be an additional duty for key, experienced members of the wing across the wing's functions who perform the above tasks when needed. Much of the work would be during crisis response and the initial creation of mission maps.

How to coordinate and manage all of these activities is the subject of the next chapter.

---

[69] We retain the term *MDT* because the MDTs largely perform this role of system defense, even though the name contains the word *mission*.

4. Managing Cyber Mission Assurance at the Wing Level

> When attempting to measure success one must make a distinction between project success and the success of the project management effort, as the two although related, may be very different.
>
> — Anton de Wit[70]

> What emerges from the studies of high reliability and resilience is a central emphasis on intensive communication and feedback, either for the individual or for and within the group which is steering the system within the boundaries of its safe envelope. It seems necessary to audit, or study in detail what it is that is taking place within this communication and feedback, in order to understand if it contains the requisite information and models to cope with the range of situations it will meet.
>
> — Andrew Hale, Frank Guldenmund, and Louis Goossens[71]

Even if all of the strategies and tasks outlined in the previous chapter are well executed, success will not be assured. For success, all personnel and units will need to work toward the same end, and leaders will need to coordinate those activities, as well as monitor activities to adjust them when needed to better align with the overall goals. Only through successful management will cyber mission assurance be achieved.

Challenges of Managing Cyber Mission Assurance

Leaders at all levels in the Department of the Air Force hierarchy face many significant challenges in managing cyber mission assurance. Two factors that complicate cyber mission assurance are the complexity of the cyber environment and the lack, to date, of actionable feedback on what works and what does not work in cyber mission assurance. These factors are interrelated.

The situation is complex because of the vast number of attack modes, because nearly any system can be attacked and, therefore, nearly every mission element is to some extent susceptible. In addition, vulnerabilities are not limited to systems but can be present in how humans behave with those systems. Humans are targeted by attackers during phishing. Humans can be insider threats. Humans sometimes fail to comply with technical orders or fail to report anomalous behavior of systems or other personnel. All of these behaviors complicate the management of cyber mission assurance.

---

[70] De Wit, 1988, p. 164.

[71] Hale, Guldenmund, and Goossens, 2006, pp. 311–312.

Concurrently, the technologies are continuously evolving. This means that the ways in which systems can be attacked change over time. What constitutes a good defense one day could be insufficient in the future. This scope of human behavior and evolving complex systems is often overwhelming to understand, monitor, and develop plans to address.

The feedback is poor relative to other threats because the threat is still emerging—adversaries have yet to impose their most exquisite cyber operations capabilities during peacetime, and their skills will no doubt mature further. The feedback that does exist suffers from being scarce, ambiguous, ephemeral, sometimes misleading, and often late—and from there being far more information on failures than on successes.[72] What level of cyber attacks will an adversary successfully launch during war? How long will the campaign continue? What effects will all of that have on systems and, ultimately, the mission? What constitutes effective resiliency to attack?

There are no definitive answers to these questions, which makes it difficult to prepare. In some sense, every system is susceptible to attack. But it is impractical to earnestly defend each system; there are too few resources for that. However, the adversary probably does not have the resources to attack everything. About the only conclusions that are clear are that no system is immune from attack and that the United States must plan for more-destructive effects during wartime than it has experienced during peacetime.

The threats from cyber attacks differ from most other threats for which a wing prepares. Such threats as are faced in air-to-air combat are fairly well known. Tactics, techniques, and procedures have been developed and are continuously reevaluated. Operators continuously rehearse how to fight in this environment. During such exercises as Red Flag, operators are put to the test. They learn from these experiences and become better at their craft. Readiness against attacks is fairly well understood and is monitored by operational readiness inspections.

For cyber mission assurance, however, comparable tactics, techniques, and procedures are in their infancy. As we heard in our wing-level interviews, wing members do not know what to reasonably expect from a cyber attack. They have had few learning experiences from which to build confidence in specific courses of action. They have not been given a clear sense of what the goal is or what constitutes an acceptable level of performance during and after cyber attacks. Consequently, monitoring is weak. Coordination mechanisms are ad hoc. These are management problems. The goal of this chapter is to offer improvements to the management of cyber mission assurance, directed at the wing level.

Dwight D. Eisenhower famously said, "Plans are worthless, but planning is everything."[73] When a real crisis happens, it will not look exactly as anticipated, which was Eisenhower's point, but the planning process should help each organization have the right tools in place and the experience in thinking through and rehearsing responses to adapt effectively. The act of planning

---

[72] For a full discussion, see Snyder et al., 2021, Chapter Two.

[73] Eisenhower, 1958, p. 818.

forces participants to confront the challenges, to identify shortfalls, to develop stratagems, and to familiarize themselves with the process of making a plan before the contingency occurs. We argue that for cyber mission assurance at the wing level, there is a need to go through the appropriate elements of a planning process for mission assurance during and after cyber attacks.

Recommendations: Wing-Level Command

The focus of management at the wing level should be those aspects that it knows best and for which the ability to adapt when under attack can only be achieved locally. What wings are best positioned to do is to monitor and to perform the day-to-day defense of systems, to respond and to recover locally from attacks, and to have plans and to exercise ways to adapt during attack to ensure mission continuity when the defense of systems fails.

Setting Goals

Wing commanders must convey a clear commander's intent to their wings. Above all, they must communicate to each member of the wing that they have a role in the cyber mission assurance for the wing and what that role is. We emphasize that this is *every* member of the wing, not just a few individuals who have specific cybersecurity duties. Cyber mission assurance is about mission assurance, not just cybersecurity. When *any* member in the wing is asked, "What is your role in the cyber mission assurance of this wing?," that person should know the answer. If the person is a member of the MDT, that role will include specific activities, such as defending a system. If the person is an aircraft maintainer, they will know that they are responsible for having a plan for when systems or data are unavailable. They will also know what constitutes suspicious activities and potentially corrupt data, and when and to whom to report these anomalies.[74]

Research shows that in complex situations in which organizations are early in the learning curve—such as cyber mission assurance—goals and oversight need to be flexible. As learning happens, the wing might need to adjust proximate goals. Initially, the commander's intent might be general and, in some cases, imprecise. With learning, the goals and commander's intent can grow to be more specific. The temptation when confronted with very difficult problems, such as cyber mission assurance, is to increase control and over-specify direction—what is sometimes called *management by direction*. But in this case, learning is most important, and the most successful style is what is sometimes called *management by discovery*.[75]

Management by discovery anticipates problems and adapts to them. Because there is a lot of uncertainty in cyber mission assurance, a commander cannot rely solely on discovering all of the vulnerabilities, identifying all of the critical systems, understanding all of the negative effects

---

[74] See Snyder et al., 2017, for an extended discussion of the role of personnel in detecting cyber anomalies.

[75] Klein, 2009, Chapter 14.

that cyber attacks could inflict on the mission, prioritizing efforts, and mitigating negative impacts. To the extent that these activities can be reliably performed, there is merit to performing them. These are indeed some of the aims of FMA-C analysis. But there are enough uncertainties in cyber mission assurance that merely doing these things can lead to overconfidence. The problem set is not well-enough known or knowable. Instead, the wing will need to continuously learn, alter goals accordingly, and always be prepared to adapt.[76] But how can such learning opportunities be created?

Creating Learning Experiences

Exactly how to make operations robust to cyber attacks is an unsolved problem. Not only do wings need to learn; so too do the Department of the Air Force and the rest of the world. Some cyber mission assurance issues can be addressed only by the wings, but, as they struggle to do so, there are limited lessons to draw upon elsewhere, as no organization has completely solved this problem. All the while, the cyber landscape continuously evolves, as do the systems and the ways in which missions are done. Given this continuous evolution, a definitive solution is unlikely. Learning will need to be continuous; wings will need to develop and sustain a learning culture for cyber mission assurance.

In the absence of real-world experience of cyber attacks in wartime, wings will need to create experiences from which their members learn.[77] As we have argued previously, to be most effective as learning experiences, artificial events—such as exercises and red-teaming—need to show participants the link between causes and effects.[78] Participants need to see how a certain action a person takes in operating a system can create a vulnerability, which can be exploited by an adversary, and how that might affect a mission. They need to see how a configuration of a system can lead to loss of data through an attack. In short, each individual needs to understand the link between cyber causes and any mission effects. Each individual needs to be able to evaluate how well they are doing their part of cyber mission assurance.[79]

Such experiences help instill the importance of sound behaviors. Wing members also need to digest learning experiences to discover new ways of accomplishing their missions when cyber defenses fail. This activity requires a big-picture view of the mission and all of the elements that combine to accomplish it. If a person in spare-parts supply identifies a work-around to losing necessary data (or confidence in data) in a data system, but that work-around is to get the data from another unit, someone needs to understand whether that other unit is equally susceptible to a common cyber attack. This need highlights the importance of dialogue across the wing,

---

[76] Klein, 2009, Chapters 14–16; Roe and Schulman, 2016.

[77] See March, Sproull, and Tamuz, 1991.

[78] See the section on feedback in Snyder et al., 2021, Chapter Two, for a full discussion.

[79] See Snyder et al., 2020, for an extended discussion.

perhaps coordinated by what we called the *teams for the ability to survive and operate in a cyber-contested environment* in the previous chapter (see Figure 3.1).[80] This team could also be the organization within the wing that memorializes learning to inculcate new members upon arrival.

To foster a learning culture for both cyber defense and cyber resiliency, the wing commander will need to encourage an atmosphere of proactivity in identifying problems and solutions and willingness to make changes as evidence indicates.[81] We have written extensively about the key attributes of an effective culture for cyber mission assurance and how leaders can instill this culture, and we refer the reader to that report.[82]

Organizations that operate at or near capacity—that is, with little slack—can still successfully achieve a learning culture, so long as the busy periods are regular in occurrence rather than clustered in time. Creating this evenness in operational tempo could at times be impossible, and, therefore, commanders should seek to use times of more-regular operational tempo to emphasize learning.[83]

Such a learning culture for cyber mission assurance might conflict with other desired cultures in a wing. A wing with a nuclear mission, for example, generally maintains a strict culture of compliance with a lesser emphasis on innovation and decentralized problem-solving. That culture is needed. Resolving this tension will be a challenge for a wing commander, and one that will need to be solved on a case-by-case basis.

Building Cyber-Ready Commanders

As with other aspects of readiness, the ultimate responsibility for cyber mission assurance in a wing lies with the wing commander. The wing commander will need to know how to use MDTs, what roles to assign to other individuals and units within the wing, and how to foster the right culture. As we have argued, cyber mission assurance is a complex, evolving problem for which there is no formulaic solution. Wing commanders will need to rely on heuristics that they have developed throughout their careers to meet this need.[84] Commanders will be better equipped and more likely to succeed if they have good heuristics for what cyber attacks can do, how cyber attacks can affect their mission(s), and what levers they have (or can create) to adjust to attacks when they occur.

---

[80] Schein, 1993, discusses the importance of cross-organizational dialogue in creating a learning culture.

[81] Schein and Schein, 2017, pp. 344–354.

[82] Snyder et al., 2021, Chapter Five.

[83] Desai, 2020.

[84] Although heuristics can lead to biases, strong evidence indicates that heuristics based on deep expertise are the best way to solve complex, dynamic problems. See Klein, 1998; Klein, 2009; and Kahneman and Klein, 2009. Some research also suggests that learning is accelerated when members are more expert; see Greenwood et al., 2019.

Such experiences have not been a necessary part of the career preparation of wing commanders. Current wing commanders have diverse experiences regarding what sophisticated cyber attack campaigns might look like and strategies to ensure missions when they occur. Some wing commanders are well equipped for this challenge, and some are not. If the Department of the Air Force truly finds cyber mission assurance to be important, it will need to emphasize career experiences that prepare wing commanders for these challenges. If not, it will continue to signal to wings that cyber mission assurance is of lesser importance than other aspects of readiness.

Although this problem can only be addressed over time, current wing commanders can help the next generation. Wing commanders can create opportunities for squadron and group commanders to develop better cyber mission assurance expertise, especially a mindset for mission assurance in the face of adversary cyber operations.

Recommendations: Above Wing Command

Wing commanders cannot accomplish these steps without direction and support from above the wing level.

Setting Goals

The foremost role of command above the wing is to set clear goals for wing-level cyber mission assurance. For wing commanders to issue an effective commander's intent, they need to receive one from their commander. Wing commanders indicated little guidance through formal command channels for how they should address cyber mission assurance or how cyber mission assurance should be prioritized over other competing wing activities, beyond general direction to ensure the wing's mission regardless of threat. To ensure that cyber mission assurance is adequately addressed, higher command needs to give wing commanders a statement of commander's intent regarding goals for cyber mission assurance.

How specific should these goals be? Guidance for specificity again comes from the theory of organizational design. Organizations managing complex situations, such as cyber mission assurance, generally decentralize control, delegating decisions to the lowest organizational level that is practical. When facing complex problems, organizations control decisionmaking by collaboration rather than standardization via fixed rules.[85] According to these principles, many of the decisions for how a wing should address cyber mission assurance should reside at the locus of essential information—the wing level—and higher command should codify guidance only to the extent that it is sure that it knows the solutions. Those solutions should be fairly stable and should not need to be frequently adjusted as situations evolve.

---

[85] Mintzberg, 1979; Snyder et al., 2015, Chapter One.

Goal specification, therefore, should not be overly prescriptive and should be restricted to what is known for sure by the level in the organization issuing the goals. Wings should be free to solve the problems for which they possess the requisite expertise. But other research shows that the more specifically goals can be stated, the more motivated directed individuals become and the more likely it is that they will be successful.[86] Higher command needs to find the most fitting stance between overly prescriptive goals and overly broad goals. The key is to be directive enough to be able to hold wing commanders accountable but to not bind wings to rules that strip them of the ability to solve problems in the new, complex, and dynamic cyber environment.

At this juncture, we recommend that, for cyber mission assurance, higher command should issue a commander's intent to wing commanders along the lines of directing them

- to defend the wing's critical systems
- to plan and exercise the ability to respond to and recover from cyber incidents
- to plan and exercise means to maintain resiliency of the wing's missions by adaptation when systems fail
- to maintain sufficient situational awareness to make decisions to accomplish defense, response and recovery, and resiliency.

Information on how to accomplish these goals does not generally reside at higher command levels, and, therefore, care should be taken to avoid over-specifying the means for how these goals should be accomplished. Higher command should allow wings to determine which systems are the most critical, how response and recovery should be done, and plans for how to adapt for mission resiliency during cyber attacks.

That is not to say that wings should not be given more-prescriptive guidance in some areas. But prescriptive guidance should come only when specific, appropriate activities are clear and the knowledge of how those activities should be executed lies outside the wing. An example from the last chapter is directives for the cyber defense of cyber-physical weapon systems. In that case, the PMO has better knowledge than the wing, and some defensive tactics, techniques, and procedures can be appropriately stated, perhaps in the form of a technical order. These decisions should not be left to the wing, but rather should be directed in appropriate specificity by organizations that possess the pertinent knowledge.

Establishing Accountability

The other component of higher command and control is accountability and the associated monitoring necessary to keep wings accountable. Wings can reasonably be held accountable only to the direction that they are given. The level of accountability, therefore, depends on the level of specificity of the goals (commander's intent), reinforcing the need for clear direction from higher command.

Wings need to be accountable in two broad areas:

---

[86] Frost and Mahoney, 1976.

- Are the directed actions being taken as envisioned? We call this facet of accountability *administrative compliance.*
- Are the actions taken by the wing achieving, or likely to achieve, the goal? We call this facet of accountability *operational effectiveness.*

These two areas are what Anton de Wit refers to as "the success of the project management effort" (administrative compliance) and "project success" (operational effectiveness) in the opening quotation for this chapter.

Administrative Compliance

Administrative compliance is the easier of the two to document and monitor. The accountability in this case is formal reporting from the wing to higher headquarters specifying how it is addressing the four goals listed earlier and the progress in carrying out those activities. Which systems is it defending and how? How were these prioritized? What are the gaps? How well are its personnel trained for this purpose? What are its plans for response and recovery? Has it exercised these plans? What plans exist for resiliency of the mission by means other than recovery of the attacked systems? How is the wing gaining situational awareness for its cyber mission assurance? Has the wing mapped the structure of its mission in a way that illuminates the most-critical systems and processes?

Operational Effectiveness

Operational effectiveness is quite difficult to monitor and, therefore, to hold a wing to account for. The only way to really know the state of cyber mission assurance of a wing is to see how well it can perform its mission(s) when under persistent cyber attack from a capable nation-state actor during wartime. That is, of course, too late. Proxies are needed.

This aspect of cyber mission assurance is in its infancy. Perhaps the best proxy for an attack is a red team. But most red teams are best equipped and trained to attack IT systems. Red teams will be needed that can attack the full range of systems, including each particular cyber-physical weapon system. They should also try to expose any poor practices of personnel, such as poor cyber hygiene.

For both administrative compliance and operational effectiveness, wings need to know that they will be regularly called to account. An effective way to do this is the main way wings are otherwise held to account, which is to incorporate cyber mission assurance into the various wing-level inspections. Operational readiness inspections are one important set. Others could include standardization and evaluation inspections, nuclear inspections, and other evaluations. Expectations in both accountability areas should be incorporated into these inspections. Higher headquarters will need to develop enough expertise and be willing to invest in the personnel and training to make this happen. If they are not, readiness for cyber mission assurance will be difficult to achieve and sustain, and higher headquarters will signal to wings that cyber mission assurance is not a priority.

Summary of Recommendations

We now summarize the principal recommendations from throughout the report, sorted by organization.

Wing-Level Recommendations

Foremost, we recommend that wing commanders take full ownership of cyber mission assurance of the wing. They must regard such resources as MDTs as tools at their disposal and direct them as needed for the wing's cyber mission assurance. They must avoid the mentality that MDTs support a "cybersecurity mission" that is distinct from the wing's mission. MDTs are tools to ensure the wing's mission. Wing commanders must also understand that MDTs alone will not provide cyber mission assurance. Every member of the wing, and every organizational unit within the wing, must play a role. Therefore, we recommend that wing commanders issue a commander's intent regarding cyber mission assurance of the wing. This commander's intent, and any subsequent commander's intent statements issued by group and squadron commanders, should assign clear cyber mission assurance role(s) to each unit of the wing. As part of this command direction, we recommend that wing commanders

- create teams for the ability to survive and operate in a cyber-contested environment who
    - perform and maintain mission maps for the wing using FMA-C or a similar method
    - recommend preemptive adjustments to mission architecture for enhanced resiliency
    - devise and exercise plans for mission continuity during and after cyber attacks
    - advise the commander, during and after cyber attacks, of potential courses of action for mission continuity.
- use MDTs exclusively for cyber defense of systems, so that they
    - defend systems, such that some team defends each system type
    - respond to cyber attacks of systems and recover system capabilities thereafter.

We further recommend that wing commanders

- create an appropriate learning culture for cyber mission assurance to solve the many problems in this area that do not have formulaic answers
- develop squadron and group commanders to have better cyber mission assurance expertise so that the next generation of wing commanders possesses better heuristics.

Above-Wing-Level Recommendations

Major Commands and Field Commands

We recommend that MAJCOMs and Field Commands establish more-centralized command and control for the response to and recovery from cyber incidents, including the following:

- clear thresholds for what constitutes a reportable incident and processes for reporting
- a command center to prioritize response, assign incidents to organizations for triage and assessment, and disseminate and compile findings.

The command center should be

- able to handle command and control for cyber incidents for IT systems, ICS, and cyber-physical weapon systems
- operational 24 hours per day, able to surge to wartime needs, and able to handle compartmented and special-access information.

We also recommend that MAJCOMs and Field Commands establish accountability for wings and deltas to evaluate wing and delta commanders on their readiness with respect to the commander's intent issued by MAJCOM and Field Commanders (see below). We recommend that this accountability cover

- administrative compliance: Are the directed actions being taken as envisioned?
- operational effectiveness: Are the actions taken by the wing achieving, or likely to achieve, the goal? This component should be included in operational readiness and other inspections to give it the prominence it needs for wing-level readiness.

To facilitate the quality and use of FMA-C analysis, we recommend that MAJCOMs and Field Commands

- provide quality assurance checks of finished FMA-C analysis products that wings perform
- create training for wing commanders on how they can best use FMA-C analysis.

We further recommend that MAJCOM and Field Command commanders issue a commander's intent to wing commanders along the lines of directing them

- to defend the wing's critical systems
- to plan and exercise the ability to respond to and recover from cyber incidents
- to plan and exercise means to maintain resiliency of the wing's missions by adaptation when systems fail
- to maintain sufficient situational awareness to make decisions to accomplish defense, response and recovery, and resiliency.

Program Management Offices

We recommend that PMOs

- develop the appropriate tools for cyber defense of cyber-physical weapon systems
- provide technical guidance to MDTs for how cyber defense of cyber-physical weapon systems ought to be performed, perhaps in the form of technical orders
- develop training for MDTs tailored to their cyber-physical weapon system.

Air University

We recommend that Air University

- enrich FMA-C training to help students move from simple classroom examples to the complicated missions of a wing
- provide exemplars or templates of sound FMA-C analysis that are similar to a wing's mission
- expand, temporarily, to provide either reachback capabilities or field teams to assist wings when they are doing mission mapping.

These recommendations and the associated offices of primary responsibility are summarized schematically in Figure 4.1.

Figure 4.1. Summary of Recommendations

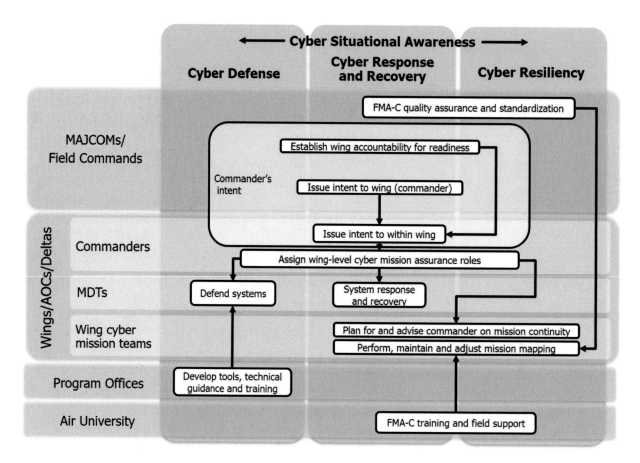

Appendix. Wing-Level Interview and Analysis Methods

In this appendix, we describe the methodology used to select, conduct, and analyze the 37 wing-level interviews. These methods were heavily informed by the tenets of the Mental Models Approach.[87] The appendix, therefore, begins with an overview of these methodological foundations and the rationale for their use.

Methodological Foundation

As discussed in the main report, the goal of these interviews was to explore how wings are (and are not) currently ensuring their cyber mission assurance and how they believe this could be improved. In performing these interviews, we hoped to reveal the current baseline, constraints, and improvements potentially knowable only to those "on the ground." Although there are many approaches for eliciting such concepts from individuals (e.g., open- or closed-ended surveys, structured interviews), most require some preexisting understanding of the current conditions and situational awareness of the respondents. At the time, previous research on such topics did not exist. We therefore chose to perform exploratory interviews, using an open-ended but semi-structured format that allowed individuals the freedom to frame the discussion in terms of how they think about the problem.

To conduct and analyze the interviews, we closely followed the interviewing method outlined in the Mental Models Approach.[88] Although the approach in its entirety was designed to develop risk communications, the preliminary steps of the approach are aimed at systematically exploring the mental model of the individuals potentially exposed to the risk in question. An individual's *mental model* is their internal representation of external reality, including their beliefs, knowledge, and values and the terms in which these are intuitively expressed.[89] The approach has been successfully applied across a number of domains, including climate risk management,[90] medical risk decisionmaking,[91] risks to aircraft sustainment efforts,[92] and flood risk.[93] By using the Mental Models Approach to interviewing and qualitative analysis, we aimed

---

[87] Morgan et al., 2002.

[88] Morgan et al., 2002.

[89] Morgan et al., 2002.

[90] For example, see Bessette et al., 2017.

[91] For example, see Haliko et al., 2018.

[92] Camm et al., 2017.

[93] Wong-Parodi, Fischhoff, and Strauss, 2018.

to understand wing commanders' and MDTs' mental models of their roles in making decisions related to cyber mission assurance.

The approach begins with conducting a set of exploratory, open-ended interviews using a funnel design, in which beginning questions ask about broad concepts and subsequent probing questions elicit more-specific detail. Use of a funnel design reduces priming effects (i.e., the interviewer presents information that could bias participant responses).[94] In its traditional form, the protocol for a mental models interview is developed based on an understanding of the mental models of individuals who are considered to be experts on the risk in question. Their *expert model* forms a normative framing of the risk—how experts understand, for example, risk exposure, consequences, and mitigations. For the purposes of our interview protocol, which is discussed in more detail later in this appendix, we relied on the extensive expertise of the project team.

Once interviews are complete, the Mental Models Approach continues with assessment of interview notes or transcriptions using a qualitative analysis. This analysis uses thematic coding,[95] in which emerging themes from the interviews are identified and delineated in a "codebook." Themes can be identified deductively, guided by the interview protocol, but are mostly developed using an inductive approach, in which emerging and prevalent concepts from the interviews are iteratively categorized and organized into a hierarchical codebook. The emerging themes are then tagged in excerpts of the interview note text. Methods for thematic coding allow qualitative text to be formulated as data for structured analysis and aid in summarizing key information from interviews. Analysis of these data themes can be used, for example, to systematically explore the prevalence and interactions of concepts and to make comparisons of these results across interviews.

It is important to note that analyses of mental models interviews are not meant to provide results that are generalizable to the entire population being studied. Instead, they are intended to provide the range of possible concepts that the population is considering and explore potential differences between populations. Following traditional practice, results from the mental models interviews are used to develop a survey that tests for larger-scale prevalence of these concepts among populations.[96] That is, this initial stage of the Mental Models Approach is a formative step to help develop theory that can later be tested.[97] Similarly, our intent in using the approach was to explore the range of beliefs, knowledge, and current conditions affecting wings' ability to

---

[94] There are many such priming effects; most extend from cognitive or motivational biases to which all humans are susceptible. For a short review of these biases and their secondary effects, see Appendix A of Mayer et al., forthcoming.

[95] Braun and Clarke, 2006.

[96] Morgan et al., 2002.

[97] Indeed, the approach is derived from tenets of grounded theory (Glaser and Strauss, 2017), an exploratory approach for developing theory grounded in qualitative data that are systematically collected and analyzed.

ensure cyber mission assurance. Any results from qualitative analyses should not be seen as generalizable to all Department of the Air Force operational wings or different types of wings.

Wing Selection

To explore the range of wings' experience with cyber mission assurance, we selected a diverse cross-section of wings and wing-level equivalents (e.g., AOCs) for interview recruitment. Given our focus on improving the role of MDTs, we began with the 84 wings that had established MDTs at the time of our recruitment. We selected 21 wings for recruitment that could provide diversity across a number of attributes, including

- MAJCOM
- principal weapon system(s)
- mission type
- whether the wing's MDT was designated by ACC as a lead MDT
- whether the principal weapon system(s) is an aircraft
- whether the weapon system operators deploy
- whether the principal weapon system(s) is an IT-networked system
- whether the wing is the host wing at its base.

Of the 21 wings invited to participate, we ultimately conducted interviews with 16. Table 2.1 provides a list of the wings interviewed, qualified according to these attributes.

Overall, our wing sample was nearly evenly split between wings with and without lead MDTs, weapon system operator deployment, networked systems as the primary weapon system, and responsibility as the host wing of the base. The sample was more heavily weighted with wings in which the primary weapon system is an aircraft (63 percent of the sample) and those in ACC (43 percent), but it did include at least one representative wing from six other MAJCOMs. Finally, the most common mission type in the sample was command and control, which was followed by ISR, fighter, and space, with representation from five other mission types. These summaries of the samples are shown in Tables A.1–A.3.

Table A.1. Wing Sample MAJCOM Representation

MAJCOM	Number	Percentage
ACC	7	43%
AFGSC	2	13%
AMC	2	13%
USSF	2	13%
AFRC	1	6%
PACAF	1	6%
USAFE	1	6%

Table A.2. Wing Sample Mission Area Representation

Mission Type	Number	Percentage
C2	4	25%
ISR	3	19%
Fighter	2	13%
Space	2	13%
Bomber	1	6%
Test/EW	1	6%
Executive airlift	1	6%
ICBM	1	6%
Air refueling	1	6%

Table A.3. Wing Sample Representation Across Other Characteristics

Wing Has the Characteristic	Lead MDT	Aircraft	Deploys	Networked System	Host Wing
Yes	50%	63%	44%	44%	50%
No	50%	37%	54%	56%	50%

Interviews

Our objective for each wing included in our sample was to conduct separate interviews with the wing commander, MDT, and, at the wing commander's discretion, any other relevant organizational element. Each interview was to cover similar topics, but the framing of interview questions and the level of detail sought were tailored to the targeted interviewee. The following section provides details on the development of the interview protocols and implementation of the interviews.

Interview Protocol

As discussed previously, we developed interview protocols using a funnel design in which general questions are posed first, followed by more-detailed probing questions based on interviewee responses, as well as topics of interest. The overarching questions and topics were developed according to the project team's current understanding of and prior research on cyber risk and resiliency, Department of the Air Force–specific cyber-related challenges, and wing-level operational considerations. Two separate protocols were developed—one for wing commanders and another for MDTs. Draft interview protocols were iteratively refined based on project team feedback and an interim quality assurance review by RAND experts and management. Interview topics covered roles and responsibilities; tasks, activities, and decisionmaking; challenges and potential improvements; information flow and organizational interactions and support; cyber situational awareness; and cyber incident processes, among others. The full protocols are included at the end of this appendix.

Figure A.1 presents an excerpt of the wing commander interview protocol. The left-hand column provides a general open-ended question, and the right-hand column provides potential probing questions and topics of interest.

Interviews Conducted

Across the 16 wings, we conducted 37 interviews: 14 with wing commanders, 13 with MDTs, and ten with other relevant wing organizational elements (e.g., Communications Squadron commanders, CDCC, Intelligence Squadron commanders). Six interviews were conducted in person, and the remaining 31 were conducted via telephone.[98] Interviews with wing commanders and other wing elements generally lasted 60 minutes, and MDT interviews varied in length between 60 and 90 minutes. Notes were captured for all interviews; these notes formed the data source for our qualitative analysis.

---

[98] In the midst of our interviews, DoD enacted travel and base visit restrictions because of the coronavirus disease of 2019 pandemic, which required all interviews after mid-March 2020 to be conducted via telephone.

Figure A.1. Excerpt of the Wing Commander Interview Protocol

HAPPENING NOW: *Currently,* what is/are *your* role(s) as the Wing commander for ensuring mission survivability in the face of adversary cyber operations?	**R&R**	Your responsibilities
		Decision authority for responsibilities (delegation?)
		Peacetime/wartime R&R differences
		Your interactions inside/outside wing (why/how often) (CDCC, CDOC, CROWS, program offices, 616th OC, 16th AF, NASIC, AO(s), AFOSI, CPTs)
	Info flow	Support/information *requests made* outside wing
		Support/information *received* inside/outside wing **(esp. cyber vulnerabilities, threats and incidents)** and use
		Information *provided* to others **(esp. up chain of command)**
	Info specific	WCC Critical Information Requirements (CCIRs)
		Regular cyber brief (separate or with group brief)
		MDT member attend daily briefing (if no, helpful?)
		Cyber information in readiness document
		Wing cyber notification requirements
OPTIMAL ROLE: What *should* the role be of the Wing Commander to ensure mission survivability in the face of adversary cyber operations?	**Shoulds**	Information needed for cyber situational awareness
		Involvement in cyber risk acceptance (e.g., incidents)
		WCC decision authority (not delegated), when?
		Decisions kicked up chain, when?
		Peacetime/wartime R&R differences
CONSTRAINTS/ IMPROVEMENTS: Are you able to effectively perform this role(s) right now? If not, why? What could be improved?	**Generic prompts**	Wing support your role (organize, train, equip)? Why not? Improvements?
		Other organizations support wing cyber-survivability? Why not? Improvements?
		Usefulness of information you receive
		Additional information you want? Why? Format
		Delegation of information you receive
		Current pursuits to improve above

NOTE: AF = Air Force; AFOSI = Air Force Office of Special Investigations; esp. = especially, R&R = roles and responsibilities; WCC = wing commander.

Qualitative Analysis

Thematic Codebook Development

The first step to performing thematic coding is to develop a *codebook*—a hierarchical list of themes that are to be identified and tagged in the interview note text. As discussed previously, themes can be identified both deductively and inductively. We developed our codebook using both approaches. We began by identifying a set of research questions that could be answered using qualitative analysis:

1. What are the primary challenges wings face to ensuring cyber mission assurance?
2. What are the enablers to wings as they work to ensure cyber mission assurance?

3. What enterprise- and/or wing-level changes could help to improve wings' cyber mission assurance?
4. What level of cyber-related interactions and support do wing commanders and MDTs have or receive from within the wing and from other organizations?
5. How do perspectives on the above questions differ between wing-level stakeholders (e.g., commanders, MDT), types of wings (e.g., different mission types), MDTs with differing characteristics (e.g., lead MDTs, size), and other wing-level characteristics? For the latter, what do these characteristics reveal about the unique aspects of wings to be considered when developing enterprise-wide improvements for cyber mission assurance?

These research questions were informed by the original project scope, ongoing research in other project tasks, and a holistic assessment of the interview discussions. The codebook was heavily informed by the interview protocols as well as the research questions. We augmented these deductive methods for codebook development through inductive exploration of emerging and prevalent themes in the interviews themselves. Accordingly, we iteratively refined the codebook as every few interviews were coded to ensure that the themes captured new relevant concepts and the appropriate level of detail.

Our resulting codebook contains three primary, "parent" codes, each with multiple "child" subcodes:

- Type of comment: Each tagged text excerpt was coded to indicate which of the five research questions it addressed (e.g., challenge, improvement).
- Theme: Each tagged excerpt was coded to indicate the theme(s) captured in that statement (e.g., the type of challenge to training or authorities).
- Stakeholder: When appropriate, each tagged excerpt was coded to indicate the stakeholders involved (e.g., lead MAJCOM or PMO).

The full codebook, including code definitions, if applicable, is provided at the end of this appendix. In addition to coding tagged excerpts, we coded a set of descriptors at the interview level. These descriptors included

- whether the interview was conducted with the wing commander, MDT, or other organizational element
- MAJCOM (ACC or other MAJCOM)
- whether the wing's MDT was designated by ACC as a lead MDT
- whether the principal weapon system(s) is an aircraft
- whether the weapon system operators deploy
- whether the principal weapon system(s) is a networked system
- whether the wing is the host wing at its base.

Thematic Coding

Coding, which was performed using Dedoose software,[99] was applied to 27 of the 37 interviews: 12 of the 13 MDT interviews, all of the wing commander interviews, and one group commander interview.[100] To ensure consistency and agreement on coded themes, as well as to further refine the codebook, two project team members individually coded seven of the interviews. After every few interviews, the team members compared codes and discussed those for which there was disagreement. Codebook definitions, inclusion criteria, and exclusion criteria were updated, and codes were added and reorganized based on these differences. Once coder agreement converged, one team member took the lead, coding the remaining 20 interviews. Areas of uncertainty for the remaining interviews were highlighted for review and discussion between the two team members. Other codes were reviewed and discussed periodically to ensure that codes were consistently applied over the course of the exercise.

Theme Prevalence to Guide Qualitative Analysis

Once thematic coding was complete, we performed simple quantitative analyses on the resulting data to guide our qualitative analysis. Our primary mode of analysis involved exploring the prevalence of coded themes or co-occurring themes (i.e., instances in which two or more codes of interest are tagged in the same text excerpt) across all interviews or within and between subsets of interviews.[101] Given that the length of interviews differed, we calculated prevalence of a theme at the interview level as the percentage of relevant coded excerpts in that interview in which that theme was coded. As an example, if an interview included 20 relevant tagged excerpts and a specific theme was coded in five of those excerpts, that theme's prevalence would be calculated as 25 percent. We followed a similar algorithm for theme co-occurrences. To determine prevalence across interviews, a simple mean prevalence was calculated. Doing so allowed for the normalization of interviews of different length. Excluding interview text that was not tagged with any code also normalized for discussions that expanded beyond the scope of our research questions. We used results of these prevalence calculations to guide our qualitative review of patterns across interviews.

---

[99] Dedoose, 2016.

[100] One MDT interview was excluded from this coding because the quality of the notes was inadequate to perform qualitative analysis. An interview with a group commander was included in this coding as a proxy for the wing commander.

[101] For analyses of specific stakeholder relationships, we used a binary indicator as opposed to prevalence. That is, if an interviewee mentioned an existing relationship once or more in an interview, that relationship was counted as existing; otherwise, the relationship was considered to not exist.

Interview Protocols

Wing Commander Interview Protocol

Overall Roles and Responsibilities (10 minutes)

Overall picture of the different cyber-related roles and responsibilities in your wing.

How does your wing ensure the survivability of the wing's missions in a cyber-contested environment, and how are these responsibilities assigned within the wing?	How	*Operating in cyber-contested environment* definition
		Fighting in cyber-contested environment and preparations
		Cyber incident response procedures
		Cyber event/incident definition
	R&R	External reach for assistance/reporting
		External organization interaction
		Peacetime/wartime R&R differences
		R&R differences across mission
		R&R for mission/installation support outside wing authority

Wing Commander Role and Support (20 minutes)

Your role specifically: (1) what is actually happening now; (2) what someone in your role should be doing; (3) why, as WCC, you can't do what you should be doing and how things can be improved.

Happening now: *Currently*, what is/are *your* role(s) as the wing commander for ensuring mission survivability in the face of adversary cyber operations?	R&R	Your responsibilities
		Decision authority for responsibilities (delegation?)
		Peacetime/wartime R&R differences
		Your interactions inside/outside wing (why/how often) (CDCC, CDOC, CROWS, program offices, 616 OC, 16th Air Force, NASIC, AO[s], AFOSI, CPTs)
	Info flow	Support/information *requests made* outside wing
		Support/information *received* inside/outside wing (especially cyber vulnerabilities, threats, and incidents) and use
		Information *provided* to others (especially up chain of command)
	Info-specific	WCC CCIRs
		Regular cyber brief (separate or with group brief)
		MDT member attends daily briefing (if no, helpful?)
		Cyber information in readiness document
		Wing cyber notification requirements
Optimal role: What *should* the role be of the wing commander to ensure mission survivability in the face of adversary cyber operations?	Shoulds	Information needed for cyber situational awareness
		Involvement in cyber risk acceptance (e.g., incidents)
		WCC decision authority (not delegated), when?
		Decisions kicked up chain, when?
		Peacetime/wartime R&R differences
Constraints/improvements: Are you able to effectively perform this role(s) right now? If not, why? What could be improved?	Generic prompts	Wing supports your role (organize, train, equip)? Why not? Improvements?
		Other organizations support wing cyber-survivability? Why not? Improvements?
		Usefulness of information you receive
		Additional information you want? Why? Format
		Delegation of information you receive

Wing Commander Role and Support (20 minutes)

	Constrain/improve	
		Current pursuits to improve above
		Organizational structure and reporting relationships
		Authorities/policy
		Culture
		Personnel/expertise
		Training
		Interactions within and outside of the wing
		Equipment/tools

Mission Defense Team Role (5 minutes)

Specifically about the MDT(s) in your wing.

What role do(es) the MDT(s) play for your wing right now? What is your commander's intent for the MDTs?	Activities	MDTs ensure survivability in cyber-contested environment
		Identify/protect mission-relevant cyber terrain
		Detect/respond/recover from cyber events
		Communicate about potential cyber events? Who, when?
		Peacetime/wartime R&R differences
		Accomplish tasks without MDT
	Manage	SQS [squadron]/MDT construct, who is developing, how?
		You issue guidance to MDTs
		MDT resource allocation (how, who, from where, future plans)

Final Questions

Is there anything else you'd like to discuss? Are there any questions that we did not ask?

Are there others you think we should speak with?

Are there documents you think would be beneficial for us to review?

Mission Defense Team Interview Protocol

Mission Defense Team Background (10 minutes/15 minutes group)

Brief overview of this MDT

What is your mission?	Bounds of mission MDT is defending
	Wing CC "commanders intent" for MDT
	MDT role in mission defense
	Wing's mission(s)
	SQS/MDT construct, who is developing, how?
What is the history of this MDT?	Length of existence; evolved (in past); if so, why?
	Current resource allocation
What are the explicit roles of MDT members?	Number of individuals in MDT
	R&R of members
	Background/skills/expertise to support R&R (specialty codes)
	Other roles in cyber squadron, how is time shared?
What training do MDT personnel receive to perform their role?	Specific to unit's mission/equipment

Mission Defense Team Activities and Tasks (20 minutes/30 minutes group)

Understand what activities and tasks the MDT performs under this role

	Identify	Identify mission-relevant cyber terrain
		Mapping of wing's mission(s), principal hazards, unacceptable losses
		Used functional mission analysis (FMA-C)
		FMA-C training, equipment (same as CPTs?)
		Looked at Minimum Essential Subsystems List
How does the MDT execute its role? What are the current tasks and activities of this MDT?	**Protect**	Protection of the cyber terrain
		Preapproved actions
	Detect	Detection of cyber events
		Monitoring procedures
	Respond	Respond to cyber events
		Cyber incident response procedures
		Cyber event/incident definition
		Analyze cyber events (assess severity, who does forensics)
		Preapproved actions
	Recover	Recovery from cyber events
		Develop/use lessons learned
	Within wing	List within wing (conditions, mechanisms, frequency)
		Directly report to?
		Wing CC interaction (content/frequency of briefings to, feedback from, and use, under what conditions)
		CCIRs
		Contribute to wing CC readiness document
Please list with whom you interact, and, for each, what is the purpose?		Your group/squadron (CC) interaction — Content/frequency of information provided and received; MDT information use
		Other group/squadron (CC) interaction
		Others in wing
	Outside wing	List outside wing (conditions, mechanisms, frequency)
		Formal reporting relationship outside wing
		Interaction with CPTs, CDOC, CDCC, CROWS, PMOs, AOs, 16th Air Force, Intelligence Community, AFOSI, other wings
		Receive and provide information (who, what type)
		Use intelligence products or data (from whom, what type, reachback for further support)
		Information about what other units discover (from whom, type—e.g., same platform, push or pull, MDT use)

Future, Constraints/Improvement (20 minutes/30 minutes group)

What is the future vision for this MDT?	**General**	Planned MDT changes (role, mission), why?
		Requirements for future plans to occur
		Changes improve wing's ability to achieve mission

		Needed changes not being pursued, why?
	Specific changes	Organizational structure and reporting relationships
		Authorities
		The constitution of the team
		Training
		Interactions and information access/reception/quality
		Funding/resource allocation
Is the MDT currently able to efficiently and effectively achieve its mission? If not, why? What constrains this?	Organize, train, equip	Level of maturity of MDT responsibilities
		Organizational structure and reporting relationships
		Authorities/policies
		Appropriate authority to connect
		Culture
		Constitution of the team (skill sets needed, experienced personnel available, resource limitations)
		Training (More needed? More-focused training [e.g., CPT] needed?)
		Equipment/tools (classification issues, contractor restrictions)
How can the MDT be improved? What would enable these improvements?	Information	Interactions and information access/reception/quality
		Actionable and trustworthy nature of information
		Intelligence products/sources (most useful, level of detail needed, example where helped perform mission)
		Access to the Joint Worldwide Intelligence Communications System? NSANet? Other sources?
		Specific information needs

Final Questions

Is there anything else you'd like to discuss? Are there any questions that we did not ask?
Are there others you think we should speak with?
Are there documents you think would be beneficial for us to review?

Thematic Codebook

Table A.4 presents the thematic codebook that we used.

Table A.4. Thematic Codebook Used for Wing-Level Interviews

Code	Definition (if applicable)
1. Type of comment	
1.1. Challenges	Circumstances or activities which are constraining wing to ensure cyber mission assurance
1.2. Enabler	Circumstances or activities which are enabling wing to ensure cyber mission assurance
1.3. Current relationship	Interactions between interviewee and other individuals or organizations
1.4. Specific improvement ideas	Concepts that are mentioned as improving cyber mission assurance or implemented activities that are implied as potential best practices

Code	Definition (if applicable)
1.5. Unique aspects of wings	Discussion of what makes a wing unique such that one-size-fits-all recommendations may not be appropriate
2. Themes	
2.1. Personnel	
2.1.1. Bodies/billets	Number of individuals or allowable individuals
2.1.2. Expertise/skills/knowledge	Individual's mastery of trainable concepts that they should have received before being stationed at the wing
2.1.3. Seniority/experience	Individual's tenure and mastery of less trainable concepts (e.g., cross-functional)
2.1.4. Attrition/continuity	Individual's ability to remain in a position or wing for sufficient time
2.1.5. Deployments	Concepts related to deployment of MDT members
2.1.6. Other	
2.2. Training	Training that should be provided or required of individuals once stationed at the wing
2.2.1. FMA-C	FMA-C training
2.2.2. Mission/system-specific	Weapon system–specific training, MQT (mission qualification training, any training relevant to the mission system, to include recurring skills training)
2.2.3. Fundamental training, IT specific	Initial Skills Training, initial qualification training—baseline training for MDT, training in Linux and Windows
2.2.4. CVA/H	Training with the MDT weapon system/toolkit, separate from MQT and IQT
2.2.5. Training timeline	Discussions related to how long training takes or how long it takes to get individuals into the training
2.2.6. Other	
2.3. Organizational structure/reporting relationships	Discussions related to how organizational structure impacts cyber mission assurance
2.3.1. Roles and responsibilities	The scope of the role and assigned activities to individuals/organizations; who does what
2.3.2. Other	
2.4. Authorities/policy	Directive policies or lack thereof
2.4.1. Classification of information	Whether information should be at a lower classification level
2.4.2. Individual authorities/jurisdiction	Whether individuals have the needed authorities and ability to make decisions
2.4.3. MDT guidance	Guidance or policy on roles and responsibilities of the MDT
2.4.4. Other	
2.5. Relationships and information flow	How information flows between, and the nature of interactions/relationships between individuals or organizations
2.5.1. Integration	The level of integration of activities, communication, etc. between individuals or organizational elements
2.5.2. Other	
2.6. Equipment/tools	Equipment and tools for the MDT
2.6.1. Acquisition process	Efficiency and effectiveness of acquiring needed equipment and tools

Code	Definition (if applicable)
2.6.2. CVA/H	CVA/H toolkit
2.6.3. Other	
2.7. Infrastructure/facilities	Physical structures and the support they provide
2.8. Resources/funding	Ability to obtain or use funding
2.9. Access	Ability to access systems or information
2.9.1. Authority to connect	Connection of tools to weapon system
2.9.2. Clearance	Ability to access classified information based on individual's clearances
2.9.3. Other	
2.10. Culture/cyber priority	Culture related to priority placed on cyber
2.11. Processes/procedures	Related to process or procedure, including both preparations/development and implementation; implementation of policy
2.11.1. FMA-C process	FMA-C-related processes
2.11.2. Other	
2.12. Other	
3. Stakeholder	
3.1. Within wing	
3.1.1. Wing CC	
3.1.2. Comm Squadron CC	
3.1.3. Intel	
3.1.4. MDT	
3.1.5. Operations	
3.1.6. Maintenance	
3.1.7. IT	
3.1.8. Group commander	
3.1.9. Other	
3.2. Outside wing	
3.2.1. PMO weapon system	
3.2.2. PMO CVA/H	
3.2.3. CROWS	
3.2.4. CDCC	
3.2.5. CDOC	
3.2.6. Intel	
3.2.7. Prime contractor	
3.2.8. Other wings	
3.2.9. Other MDTs	
3.2.10. AOs	
3.2.11. CYBERCOM/16 AF/616 OC	
3.2.12. CPTs	

Code	Definition (if applicable)
3.2.13. Parent MAJCOM	
3.2.14. Lead MAJCOM	
3.2.15. Training entities	
3.2.16. Other services	
3.2.17. HAF	
3.2.18. Other	

NOTE: AF = Air Force; CYBERCOM = U.S. Cyber Command; HAF = Headquarters U.S. Air Force; IQT = Initial Qualification Training.

References

16th Air Force, *16th Air Force (Air Forces Cyber) Capabilities, Products, and Services Catalog*, CDRL A103, March 2020.

Air Combat Command, *Cyber Defense Coordination Center (CDCC) Operating Concept*, January 26, 2018.

———, *Air Combat Command as Air Force Lead Command for Cyber Forces: Air Force Mission Defense Team (MDT) Operating Concept*, January 2020, Not available to the general public.

Air Force Instruction 13-1AOC, Vol. 3, *Operational Procedures—Air Operations Center (AOC)*, Washington, D.C.: Department of the Air Force, November 2, 2011, Incorporating Change 1, May 18, 2012.

Air Force Instruction 17-101, *Risk Management Framework (RMF) for Air Force Information Technology (IT)*, Washington, D.C.: Department of the Air Force, February 6, 2020.

Air Force Instruction 17-203, *Cyber Incident Handling*, Washington, D.C.: Department of the Air Force, March 16, 2017.

Air Force Instruction 38-101, *Manpower and Organization*, Washington, D.C.: Department of the Air Force, August 29, 2019.

Alur, Rajeev, *Principles of Cyber-Physical Systems*, Cambridge, Mass.: The MIT Press, 2015.

Ashibani, Yosef, and Qusay H. Mahmoud, "Cyber Physical Systems Security: Analysis, Challenges and Solutions," *Computers & Security*, Vol. 68, July 2017, pp. 81–97.

Bartock, Michael, Jeffrey Cichonski, Murugiah Souppaya, Matthew Smith, Greg Witte, and Karen Scarfone, *Guide for Cybersecurity Event Recovery*, Washington, D.C.: U.S. Department of Commerce, NIST Special Publication 800-184, December 2016.

Bell, Gerald D., "Determinants of Span of Control," *American Journal of Sociology*, Vol. 73, No. 1, July 1967, pp. 100–109.

Bessette, Douglas L., Lauren A. Mayer, Bryan Cwik, Martin Vezér, Klaus Keller, Robert J. Lempert, and Nancy Tuana, "Building a Values-Informed Mental Model for New Orleans Climate Risk Management," *Risk Analysis*, Vol. 37, No. 10, October 2017, pp. 1993–2004.

Bhamare, Deval, Maede Zolanvari, Aiman Erbad, Raj Jain, Khaled Khan, and Nader Meskin, "Cybersecurity for Industrial Control Systems: A Survey," *Computers & Security*, Vol. 89, February 2020.

Blau, Peter M., "The Hierarchy of Authority in Organizations," *American Journal of Sociology*, Vol. 73, No. 4, January 1968, pp. 453–467.

Braun, Virginia, and Victoria Clarke, "Using Thematic Analysis in Psychology," *Qualitative Research in Psychology*, Vol. 3, No. 2, 2006, pp. 77–101.

Bryant, William D., "Mission Assurance Through Integrated Cyber Defense," *Air & Space Power Journal*, Vol. 30, No. 4, Winter 2016, pp. 5–17.

Buchanan, Ben, *The Hacker and the State: Cyber Attacks and the New Normal of Geopolitics*, Cambridge, Mass.: Harvard University Press, 2020.

Camm, Frank, John Matsumura, Lauren A. Mayer, and Kyle Siler-Evans, *A New Methodology for Conducting Product Support Business Case Analysis (BCA): With Illustrations from the F-22 Product Support BCA*, Santa Monica, Calif.: RAND Corporation, RR-1664-AF, 2017. As of June 1, 2021:
https://www.rand.org/pubs/research_reports/RR1664.html

Carter, Joshua, "552nd Air Control Networks Squadron Creates First Qualification Training for Mission Defense Teams," U.S. Air Force, November 12, 2019.

Chairman of the Joint Chiefs of Staff Manual 3150.07E, *Joint Reporting Structure for Cyberspace Operations Status*, Washington, D.C., November 8, 2013.

Chairman of the Joint Chiefs of Staff Manual 6510.01B, *Cyber Incident Handling Program*, Washington, D.C., July 10, 2012, Directive Current as of December 18, 2014.

Cichonski, Paul, Tom Millar, Tim Grance, and Karen Scarfone, *Computer Security Incident Handling Guide: Recommendations of the National Institute of Standards and Technology*, Washington, D.C.: U.S. Department of Commerce, NIST Special Publication 800-61, Revision 2, August 2012.

Coats, Jennifer C., and Frederick W. Rankin, "The Delegation of Decision Rights: An Experimental Investigation," in Marc J. Epstein and Mary A. Malina, eds., *Advances in Management Accounting*, Vol. 27, Somerville, Mass.: Emerald Publishing, 2017, pp. 39–71.

Cohen, Rachel S., "16th Air Force Launches Information Ops for the Digital Age," *Air Force Magazine*, December 2019, pp. 33–36.

Committee on National Security Systems, *Committee on National Security Systems (CNSS) Glossary*, CNSSI No. 4009, April 6, 2015.

———, *Cyber Incident Response*, CNSSI 1010, December 2016, Not available to the general public.

Corey, Juliane, and Mikaela Strobel, "Hanscom Begins Network-as-a-Service Experiments," Air Force Life Cycle Management Center, September 27, 2018.

Cyber Resiliency Steering Group, *Air Force Cyber Resiliency Office for Weapon Systems (CROWS) Cyber Incident Coordination Cell (CICC) and Cyber Incident Response Team (IRT) for Weapon Systems Concept of Operations (CONOPS)*, September 15, 2016, Not available to the general public.

Danzig, Richard J., *Surviving on a Diet of Poisoned Fruit: Reducing the National Security Risks of America's Cyber Dependencies*, Washington, D.C.: Center for New American Security, July 2014.

Dedoose, web application for managing, analyzing, and presenting qualitative and mixed-methods research data, Los Angeles, Calif.: SocioCultural Research Consultants, LLC, Version 7.0.23, 2016. As of June 4, 2021:
www.dedoose.com

Desai, Vinit M., "Can Busy Organizations Learn to Get Better? Distinguishing Between the Competing Effects of Constrained Capacity on the Organizational Learning Process," *Organization Science*, Vol. 31, No. 1, January–February 2020, pp. 67–84.

Dessein, Wouter, "Authority and Communication in Organizations," *Review of Economic Studies*, Vol. 69, No. 4, October 2002, pp. 811–838.

de Wit, Anton, "Measurement of Project Success," *International Journal of Project Management*, Vol. 6, No. 3, August 1988, pp. 164–170.

Dobrajska, Magdalena, Stephan Billinger, and Samina Karim, "Delegation Within Hierarchies: How Information Processing and Knowledge Characteristics Influence the Allocation of Formal and Real Decision Authority," *Organization Science*, Vol. 26, No. 3, May–June 2015, pp. 687–704.

DoD—*See* U.S. Department of Defense.

Eisenhower, Dwight D., "Remarks at the National Defense Executive Reserve Conference," in *Public Papers of the Presidents of the United States: Dwight D. Eisenhower, 1957, Containing the Public Messages, Speeches, and Statements of the President, January 1 to December 31, 1957*, Washington, D.C.: Office of the Federal Register, National Archives and Records Service, 1958.

Freedman, Lawrence, *Strategy: A History*, New York: Oxford University Press, 2013.

Frost, Peter J., and Thomas A. Mahoney, "Goal Setting and the Task Process: I. An Interactive Influence on Individual Performance," *Organizational Behavior and Human Performance*, Vol. 17, No. 2, December 1976, pp. 328–350.

Gaim, Medhanie, Nils Wåhlin, Miguel Pina e Cunha, and Stewart Clegg, "Analyzing Competing Demands in Organizations: A Systematic Comparison," *Journal of Organization Design*, Vol. 7, 2018.

Glaser, Barney G., and Anselm L. Strauss, *Discovery of Grounded Theory: Strategies for Qualitative Research*, New York: Routledge, 2017.

Gonda, Oded, "Understanding the Threat to SCADA Networks," *Network Security*, Vol. 2014, No. 9, September 2014, pp. 17–18.

Gosler, James R., and Lewis Von Thaer, *Resilient Military Systems and the Advanced Cyber Threat*, Washington, D.C.: U.S. Department of Defense, Defense Science Board, Task Force Report, January 2013.

Greenberg, Andy, "The Code That Crashed the World: The Untold Story of NotPetya, the Most Devastating Cyberattack in History," *Wired*, September 2018, pp. 52–63.

Greenwood, Brad N., Ritu Agarwal, Rajshree Agarwal, and Anandasivam Gopal, "The Role of Individual and Organizational Expertise in the Adoption of New Practices," *Organization Science*, Vol. 30, No. 1, January–February 2019, pp. 191–213.

Greer, Christopher, Martin Burns, David Wollman, and Edward Griffor, *Cyber-Physical Systems and Internet of Things*, Washington, D.C.: U.S. Department of Commerce, NIST Special Publication 1900-202, March 2019.

Hagen, Jeff, Lillian Ablon, John Ausink, Jeffrey Brown, Michael Decker, Quentin Hodson, Myron Jura, Mary Lee, and Brynn Tannehill, *Piecing Together the Puzzle: Building Cyber Resilient Air Force Missions*, Santa Monica, Calif.: RAND Corporation, 2021, Not available to the general public.

Hale, Andrew, Frank Guldenmund, and Louis Goossens, "Auditing Resilience in Risk Control and Safety Management Systems," in Erik Hollnagel, David D. Woods, and Nancy Leveson, eds., *Resilience Engineering: Concepts and Precepts*, Burlington, Vt.: Ashgate Publishing Company, 2006, pp. 289–314.

Haliko, Shannon, Julie Downs, Deepika Mohan, Robert Arnold, and Amber E. Barnato, "Hospital-Based Physicians' Intubation Decisions and Associated Mental Models When Managing a Critically and Terminally Ill Older Patient," *Medical Decision Making*, Vol. 38, No. 3, April 2018, pp. 344–354.

Headquarters United States Air Force (HQ USAF), *Implementation of Department of the Air Force Cyber Squadrons*, Washington, D.C.: Department of the Air Force, Program Action Directive (PAD) D15-03, May 12, 2020, Not available to the general public.

HQ USAF/A4, "Civil Engineer Control Systems Cybersecurity," Washington, D.C.: Department of the Air Force, Air Force Guidance Memorandum AFGM2019-32-02, September 5, 2019.

Jensen, Michael C., and William H. Meckling, "Specific and General Knowledge and Organizational Structure," in Lars Werin and Han Wijkander, eds., *Contract Economics*, Cambridge, Mass.: Basil Blackwell, 1992, pp. 251–274.

Joint Publication 2-0, *Joint Intelligence*, Washington, D.C.: Chairman of the Joint Chiefs of Staff, October 22, 2013.

Joint Publication 3-0, *Joint Operations*, Washington, D.C.: Chairman of the Joint Chiefs of Staff, January 17, 2017, Incorporating Change 1, October 22, 2018.

Joint Publication 3-12, *Cyberspace Operations*, Washington, D.C.: Chairman of the Joint Chiefs of Staff, June 8, 2018.

Kahneman, Daniel, *Thinking, Fast and Slow*, New York: Farrar, Straus and Giroux, 2011.

Kahneman, Daniel, and Gary Klein, "Conditions for Intuitive Expertise: A Failure to Disagree," *American Psychologist*, Vol. 64, No. 6, September 2009, pp. 515–526.

Kent, Glenn A., with David Ochmanek, Michael Spirtas, and Bruce R. Pirnie, *Thinking About America's Defense: An Analytical Memoir*, Santa Monica, Calif.: RAND Corporation, OP-223-AF, 2008. As of June 1, 2021:
https://www.rand.org/pubs/occasional_papers/OP223.html

King, Angus, and Mike Gallagher, *Cyberspace Solarium Commission*, Washington, D.C.: U.S. Cyberspace Solarium Commission, March 2020.

Kirby, Lynn, "USSF Field Command Structure Reduces Command Layers, Focuses on Space Warfighter Needs," United States Space Force Public Affairs, June 30, 2020.

Klein, Gary, *Sources of Power: How People Make Decisions*, Cambridge, Mass.: The MIT Press, 1998.

———, *Streetlights and Shadows: Searching for the Keys to Adaptive Decision Making*, Cambridge, Mass.: The MIT Press, 2009.

Knowles, William, Daniel Prince, David Hutchison, Jules Ferdinand Pagna Disso, and Kevin Jones, "A Survey of Cybersecurity Management in Industrial Control Systems," *International Journal of Critical Infrastructure Protection*, Vol. 9, June 2015, pp. 52–80.

Langner, Ralph, *To Kill a Centrifuge: A Technical Analysis of What Stuxnet's Creators Tried to Achieve*, Arlington, Va.: The Langner Group, 2013.

Leveson, Nancy, "A New Accident Model for Engineering Safer Systems," *Safety Science*, Vol. 42, No. 4, 2004, pp. 237–270.

———, *Engineering a Safer World: Systems Thinking Applied to Safety*, Cambridge, Mass.: The MIT Press, 2011.

———, "A Systems Approach to Risk Management Through Leading Safety Indicators," *Reliability Engineering & System Safety*, Vol. 136, April 2015, pp. 17–34.

Leveson, Nancy, Nicolas Dulac, David Zipkin, Joel Cutcher-Gershenfeld, John Carroll, and Betty Barrett, "Engineering Resilience into Safety-Critical Systems," in Erik Hollnagel, David D. Woods, and Nancy Leveson, eds., *Resilience Engineering: Concepts and Precepts*, Burlington, Vt.: Ashgate Publishing Company, 2006, pp. 95–123.

March, James G., Lee S. Sproull, and Michal Tamuz, "Learning from Samples of One or Fewer," *Organization Science*, Vol. 2, No. 1, February 1991, pp. 1–13.

Marino, Anthony M., and John G. Matsusaka, "Decision Processes, Agency Problems, and Information: An Economic Analysis of Capital Budgeting Procedures, *Review of Financial Studies*, Vol. 18, No. 1, Spring 2005, pp. 301–325.

Mayer, Lauren A., Don Snyder, Guy Weichenberg, Danielle C. Tarraf, Jonathan William Welburn, Suzanne Genc, Myron Hura, and Bernard Fox, *Cyber Mission Thread Analysis: An Implementation Guide for Process Planning and Execution*, Santa Monica, Calif.: RAND Corporation, RR-3188/2-AF, forthcoming.

McCarthy, James, Michael Powell, Keith Stouffer, CheeYee Tang, Timothy Zimmerman, William Barker, Titilayo Ogunyale, Devin Wynne, and Johnathan Wiltberger, *Securing Manufacturing Industrial Control Systems: Behavioral Anomaly Detection*, Washington, D.C.: U.S. Department of Commerce, NISTIR 8219, July 2020.

Mintzberg, Henry, *The Structuring of Organizations: A Synthesis of Research*, Englewood Cliffs, N.J.: Prentice-Hall, 1979.

Morgan, M. Granger, "Use (and Abuse) of Expert Elicitation in Support of Decision Making for Public Policy," *PNAS*, Vol. 111, No. 20, May 20, 2014, pp. 7176–7184.

Morgan, M. Granger, Baruch Fischhoff, Ann Bostrom, and Cynthia J. Atman, *Risk Communication: A Mental Models Approach*, Cambridge, United Kingdom: Cambridge University Press, 2002.

Naderifar, Mahin, Hamideh Goli, and Fereshteh Ghaljaie, "Snowball Sampling: A Purposeful Method of Sampling in Qualitative Research," *Strides in Development of Medical Education*, Vol. 14, No. 3, September 2017.

Nazir, Sajid, Shushma Patel, and Dilip Patel, "Assessing and Augmenting SCADA Cyber Security: A Survey of Techniques," *Computers & Security*, Vol. 70, 2017, pp. 436–454.

Nourian, Arash, and Stuart Madnick, "A Systems Theoretic Approach to the Security Threats in Cyber Physical Systems Applied to Stuxnet," *IEEE Transactions on Dependable and Secure Computing*, Vol. 15, No. 1, January–February 2018, pp. 2–13.

Ouchi, William G., and John B. Dowling, "Defining the Span of Control," *Administrative Science Quarterly*, Vol. 19, No. 3, September 1974, pp. 357–365.

Recorded Future, homepage, undated. As of June 4, 2021:
https://www.recordedfuture.com/

Roe, Emery, and Paul R. Schulman, *Reliability and Risk: The Challenge of Managing Interconnected Infrastructures*, Stanford, Calif.: Stanford University Press, 2016.

Sanger, David E., *The Perfect Weapon: War, Sabotage, and Fear in the Cyber Age*, New York: Crown, 2018.

Schein, Edgar H., "On Dialogue, Culture, and Organizational Learning," *Organizational Dynamics*, Vol. 22, No. 2, Autumn 1993, pp. 40–51.

Schein, Edgar H., with Peter Schein, *Organizational Culture and Leadership*, 5th ed., Hoboken, N.J.: John Wiley & Sons, Inc., 2017.

Secretary of the Air Force, *Maintenance Assistance*, Washington, D.C., TO 00-25-107, August 15, 2011.

Snyder, Don, Elizabeth Bodine-Baron, Mahyar A. Amouzegar, Kristin F. Lynch, Mary Lee, and John G. Drew, *Robust and Resilient Logistics Operations in a Degraded Information Environment*, Santa Monica, Calif.: RAND Corporation, RR-2015-AF, 2017. As of June 2, 2021:
https://www.rand.org/pubs/research_reports/RR2015.html

Snyder, Don, Elizabeth Bodine-Baron, Dahlia Anne Goldfeld, Bernard Fox, Myron Hura, Mahyar A. Amouzegar, and Lauren Kendrick, *Cyber Mission Thread Analysis: A Prototype Framework for Assessing Impact to Missions from Cyber Attacks to Weapon Systems*, Santa Monica, Calif.: RAND Corporation, RR-3188/1-AF, forthcoming.

Snyder, Don, Lauren A. Mayer, Myron Hura, Suzanne Genc, Colby Peyton Steiner, Laura Werber, Kathryn O'Connor, Keith Gierlack, Paul Dreyer, and Bernard Fox, *Managing for Mission Assurance in the Face of Advanced Cyber Threats*, Santa Monica, Calif.: RAND Corporation, RR-4198-AF, 2021. As of July 8, 2021:
https://www.rand.org/pubs/research_reports/RR4198.html

Snyder, Don, Lauren A. Mayer, Guy Weichenberg, Danielle C. Tarraf, Bernard Fox, Myron Hura, Suzanne Genc, and Jonathan William Welburn, *Measuring Cybersecurity and Cyber Resiliency*, Santa Monica, Calif.: RAND Corporation, RR-2703-AF, 2020. As of June 2, 2021:
https://www.rand.org/pubs/research_reports/RR2703.html

Snyder, Don, James D. Powers, Elizabeth Bodine-Baron, Bernard Fox, Lauren Kendrick, and Michael H. Powell, *Improving the Cybersecurity of U.S. Air Force Military Systems Throughout Their Life Cycles*, Santa Monica, Calif.: RAND Corporation, RR-1007-AF, 2015. As of June 2, 2021:
https://www.rand.org/pubs/research_reports/RR1007.html

Upadhyay, Darshana, and Srinivas Sampalli, "SCADA (Supervisory Control and Data Acquisition) Systems: Vulnerability Assessment and Security Recommendations," *Computers & Security*, Vol. 89, February 2020.

U.S. Air Force Doctrine, *Air Force Glossary*, Curtis E. LeMay Center for Doctrine and Education, November 3, 2020. As of December 10, 2020:
https://www.doctrine.af.mil/External-Links/Air-Force-Glossary/

U.S. Department of Defense, *Department of Defense Cyber Strategy: Summary*, Washington, D.C., 2018.

U.S. Government Accountability Office, *Weapon Systems Cybersecurity: DOD Just Beginning to Grapple with Scale of Vulnerabilities*, Washington, D.C., GAO-19-128, October 2018.

Valentine, W. Mark, *Organizing for Cyber Resilience: Rethinking the Balance Between Prevention and Response*, Arlington, Va.: Mitchell Institute for Aerospace Studies, The Mitchell Forum No. 20, October 2018.

The White House, *Presidential Policy Directive—United States Cyber Incident Coordination*, Washington, D.C., PPD-41, July 26, 2016.

Wong-Parodi, Gabrielle, Baruch Fischhoff, and Benjamin Strauss, "Effect of Risk and Protective Decision Aids on Flood Preparation in Vulnerable Communities," *Weather, Climate, and Society*, Vol. 10, No. 3, July 2018, pp. 401–417.

Young, William, and Nancy G. Leveson, "An Integrated Approach to Safety and Security Based on Systems Theory," *Communications of the ACM*, Vol. 57, No. 2, February 2014, pp. 31–35.